Lecture Notes in Electrical Engineering

Volume 400

About this Series

"Lecture Notes in Electrical Engineering (LNEE)" is a book series which reports the latest research and developments in Electrical Engineering, namely:

- Communication, Networks, and Information Theory
- Computer Engineering
- Signal, Image, Speech and Information Processing
- Circuits and Systems
- Bioengineering

LNEE publishes authored monographs and contributed volumes which present cutting edge research information as well as new perspectives on classical fields, while maintaining Springer's high standards of academic excellence. Also considered for publication are lecture materials, proceedings, and other related materials of exceptionally high quality and interest. The subject matter should be original and timely, reporting the latest research and developments in all areas of electrical engineering.

The audience for the books in LNEE consists of advanced level students, researchers, and industry professionals working at the forefront of their fields. Much like Springer's other Lecture Notes series, LNEE will be distributed through Springer's print and electronic publishing channels.

More information about this series at http://www.springer.com/series/7818

Yang Xu
Editor

Nuclear Power Plants: Innovative Technologies for Instrumentation and Control Systems

International Symposium on Software Reliability, Industrial Safety, Cyber Security and Physical Protection of Nuclear Power Plant

Springer

Editor
Yang Xu
Department of Engineering Physics
Tsinghua University
Beijing
China

ISSN 1876-1100 ISSN 1876-1119 (electronic)
Lecture Notes in Electrical Engineering
ISBN 978-981-10-3360-5 ISBN 978-981-10-3361-2 (eBook)
DOI 10.1007/978-981-10-3361-2

Library of Congress Control Number: 2016963309

Printed on acid-free paper

This Springer imprint is published by Springer Nature
The registered company is Springer Nature Singapore Pte Ltd.
The registered company address is: 152 Beach Road, #21-01/04 Gateway East, Singapore 189721, Singapore

Preface

The International Symposium on Software Reliability, Industrial Safety, Cyber Security and Physical Protection of Nuclear Power Plant was successfully held in Yinchuan, China, from May 30 to June 2, 2016. This symposium was jointly organized by China Instrument and Control Society and the Government of Yinchuan City.

The purpose of this symposium is to discuss the status quo, technical progress and development direction of digital instrument control technology, software reliability, information security and physical protection in the process of nuclear power development. It aims to provide a platform of technical exchange and experience sharing for those broad masses of experts and scholars and nuclear power practitioners. At the same time, it also provides a platform for the combination of production, teaching and research in universities and enterprises to promote the safe development of nuclear power plant.

This symposium received a total of 50 papers, covering digital instrument control technology, software reliability, information security and physical protection and other technical directions. After an anonymous peer review and expert selection, 24 excellent papers were finally recommended to Springer for publication. These 24 papers presented corresponding technical solutions and new ideas of the main design of the nuclear power plant control system, software verification and validation, digital instrument control system CPU redundancy configuration, fault diagnosis, remote I/O technology and software reliability, etc. We definitely believe that these papers in this volume have the innovative and practical value, and they can provide a useful reference for the follow-up-related work of nuclear power plant.

201 experts and scholars, students, corporate executives and outstanding scientist and technological workers from 68 institutes and organizations attended this symposium, including Tsinghua University, the Ministry of Environmental Protection, the Chinese Academy of Sciences and the China General Nuclear Power Group. Mr. Wang Zhongqiu, director of Nuclear and Radiation Safety Review Center of Environmental Protection Department, gives the keynote speech of "Nuclear power plant digital instrument control system review", and Mr. Chen

Weihua, director of Instrumentation and Control Department of China Guangdong Nuclear Power Design Corporation, gives the invited lecture of "Discussion on Information Security of Digital Instrument Control System in Nuclear Power Plant". Many other wonderful presentations were also presented by experts such as Prof. Huang Xiaojin, researcher of Tsinghua University, Mr. Gao Feng, vice president of China Guangdong Nuclear Power Design Corporation and other industry professionals. A wide ranging discussion were held in the final section of this symposium, which focus on the safety review and information security and other content people generally concerned about.

The symposium was a great success, and we must give our thanks to the organizers, to the scientific society, external reviewers, subject report experts and session chairs, also to the authors of these papers, and so on. Thanks again to the great effort of each of them. Prof. Yang XuTsinghua University, Beijing, China.

Beijing, China Yang Xu

Contents

3D Digital Virtual Simulation Application System for Technical Management of Nuclear Power Station Equipment

Yi-Yang Xing, Hua-Bo Huang and Xiao-Ming Yang

Abstract 3D digital virtual simulation application system is designed to improve the risk prevention capability, reduce the maintenance cost and strengthen the refined technical management of nuclear power station equipment. By using virtual reality technology, the assembly and disassembly process, maintenance process and operation principle of nuclear power station equipment can be displayed and simulated in the system. And with the effective association between 3D model, operation data and maintenance data, the system can help professionals to improve and standardize the management of maintenance, quality control and engineering log. This paper expounds the application value and function of the 3D digital virtual simulation of nuclear power station, and presents the technical route of the whole software system.

Keywords Nuclear power station · Technical management of equipment · Virtual reality · 3D · Simulation application

1 Introduction

According to the 13th Five-Year Plan in China, the installed capacity of operational nuclear power station will reach 58 million kW, and those of nuclear power station under construction will reach 30 million kW in 2020, which will increase the nuclear power generation ratio from 2 to 4%. It is expected that the nuclear power generation ratio will be increased to 8–10% till 2030, which indicates nuclear power is an important direction for the development of Chinese energy structure optimization. However, due to the particularity of production object and process of nuclear power, safety problem has attracted a lot of attention. Except the safety

Y.-Y. Xing · H.-B. Huang (✉) · X.-M. Yang
Nanjing Isane Information Technology Co., Ltd., No. 6, Jiangjun Avenue, Jiangning District, Nanjing City 210000, Jiangsu Province, China
e-mail: hhb@isane.com.cn

Y. Xu (ed.), *Nuclear Power Plants: Innovative Technologies for Instrumentation and Control Systems*, Lecture Notes in Electrical Engineering 400,
DOI 10.1007/978-981-10-3361-2_1

measures in the design of reactor, excellent equipment technical management of professionals is also an important factor in ensuring the safety of nuclear power plant.

Equipment maintenance is important work of nuclear power station equipment technical management. There are two kinds of maintenance: one is to ensure the normal operation of system and equipment as designed, and the other one is to repair the failure system and equipment to restore the operational ability as designed. The maintenance of nuclear power station is different from those of the conventional power station or other industrial plants for its large scale, complex equipment, big loss of outage, its importance in nuclear safety and public health protection. Therefore, the technical requirements of equipment maintenance of nuclear power station is very high. The short-term task of nuclear power station maintenance department is to repair the failure equipment as soon as possible to ensure the secure and reliable operation of the units, and the long-term task is to analyze the failure, find out the countermeasures, formulate and improve the practical maintenance plan, and implement the plan.

By using virtual reality technology, combined with database, software, network and graphics technology, 3D digital virtual simulation application system of nuclear power station equipment technical management can establish a 3D digital equipment technical management platform with integrated services, data information and 3D visualization. The 3D display and virtual simulation of equipment structure, disassembly and assembly process of equipment, maintenance process and operating principle can be achieved through this platform, also a comprehensive and effective association among the online operation data, historical maintenance data and equipment basic information can be created in the 3D environment. Therefore, for the equipment technical management with core work of maintenance, both short-term and long-term task, the professionals can understand equipment location, structure, environment, current status and historical maintenance data through the platform in real-time, which helps to formulate quickly and effectively the construction and maintenance, quality control, debugging and commissioning, acceptance check plan and the allocation of related resources, and provides the 3D visualization means for nuclear power station maintenance safety, standardization and speedy and thus to enhance refined equipment technical management and improve the risk prevention capacity of nuclear power station.

2 System Design

A B/S architecture is used in nuclear power station equipment technical management virtual simulation application software. The application function and database are deployed in the back server. Client of nuclear power station network can access the system through web browser, and operate the corresponding function according to the operating authority.

2.1 System Function Design

The 3D digital virtual simulation application software of nuclear power station equipment technical management includes five parts: 3D visualization model, 3D equipment log, equipment maintenance process simulation, equipment operation principle simulation, and equipment operation parameters in real-time 3D display. First, set up visualized models for nuclear power station and all equipment, then develop integrated application software for 3D models and equipment technical management logic, database, and interactive operation, forming a 3D digitized virtual simulation platform, achieving 3D virtual simulation and refined technical management implementation of each equipment structure, principle and maintenance process.

Nuclear power plant equipment management level and technical proficiency is one of the important factors to guarantee the safety of the nuclear power plant, the 3D simulation system of nuclear power plant has a realization of a simple display device or roaming [1]. The system based on virtual reality technology with equipment management business and database data form a complete set of virtual simulation system, and ERP and EAM information systems and docking, can be used for guiding equipment maintenance support sqlserver, Oracle and other mainstream relational database (Fig. 1).

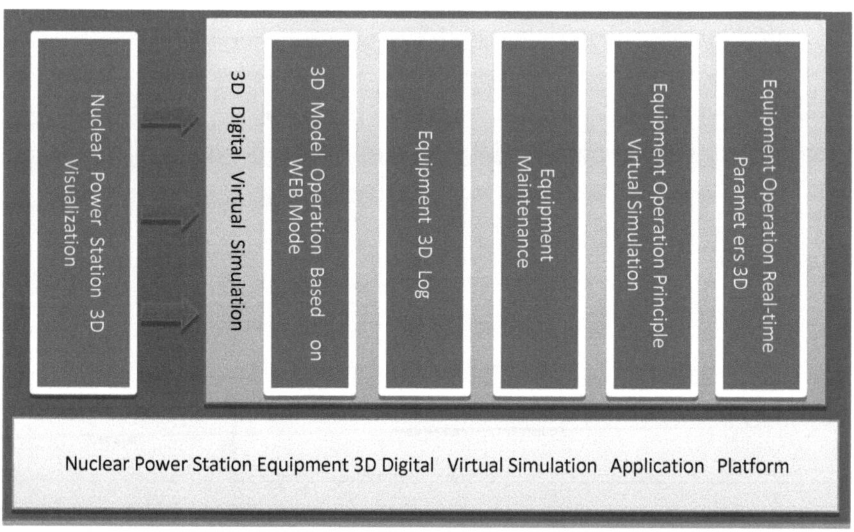

Fig. 1 System function structure

2.2 Technical Architecture

The technical architecture of 3D digital virtual simulation application system of nuclear power station equipment technical management includes the core 3D digital engine, 3D scene editing, IDE integrated development environment and 3D display.

1. Resolve equipment 3D model into 3D image through graphics rendering of 3D engine, the rendering effect is related to the equipment material.
2. Achieve roaming function of 3D scene through scene editing, 3D coordinate and angle of camera controlling and other development of technical parameters.
3. Simulate the wind, steam, smoke, water and other effects in real world through controlling the particle emission rate, emission direction, particle density and other parameters of the particle system in 3D engine by mathematical calculation.
4. Read and modify the related equipment information in relational database through data exchange module to realize the integration of management information system and 3D digital virtual simulation application system.
5. The system support iOS, Android, Window and Mac operating systems, and support web browsers (Fig. 2).

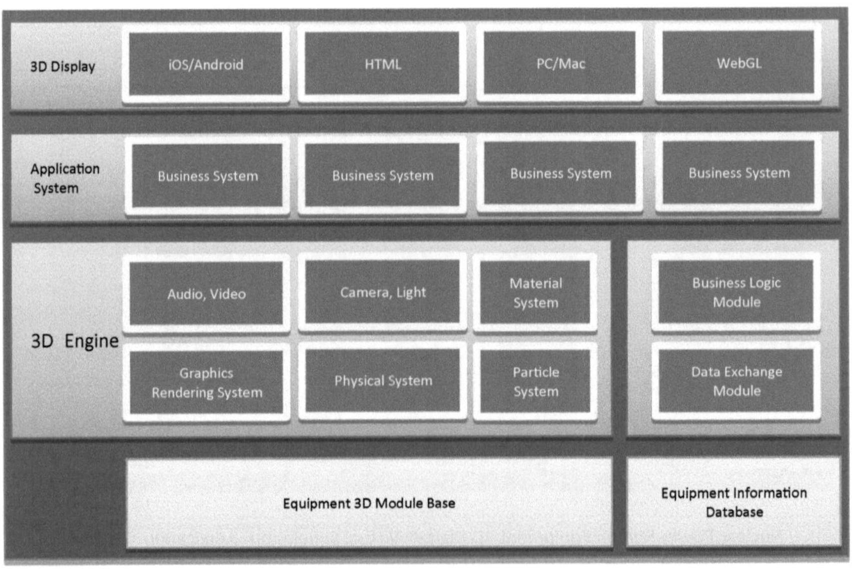

Fig. 2 System technical structure

3 3D Modeling

Nuclear power station is composed of three parts: nuclear island, conventional island and auxiliary system [2]. The physical part includes buildings, systems and equipment. Nuclear power station 3D modeling is divided into plant modeling and equipment 3D modeling. Plant modeling is for understanding plant appearance and nuclear power station layout by scene roaming; and equipment 3D modeling is the base of 3D virtual simulation system application with the scale of 1:1 to the real dimensions of equipment structure, and with the basic modeling primitives of equipment components.

The 3D modeling tool is 3dsMax. The total number and number of surface of models must be fully considered during the modeling. And the model compression, instantiation, mapping and other technical optimization need to be implemented after 3D model is built to ensure the smoothness, stability and realness in the process of using the application system.

4 System Application Function

4.1 *The Operation of 3D Model Based on WEB Mode*

The operation of 3D model based on WEB mode includes scene roaming and equipment interactive operation.

Different directions of 3D virtual scene is displayed by translation, rotation, push and pull, shake, zoom and combined above methods of camera [3]. The mouse and keyboard can be used as roaming controller and roam freely in the scene, a fixed line can also be set for scene roaming to get best effect of 3D virtual scene.

Equipment interactive operation is the operation of 3D equipment model based on WEB environment, including positioning, zooming, rotation, hiding, perspective, separate display, etc.

4.2 *Equipment 3D Log*

Nuclear power station equipment 3D log integrates the 1D data, 2D drawings and 3D structure information of equipment together, building the comprehensive, visualized equipment archive, which helps accessing information and review data conveniently and mastering equipment structure, service status at any time.

Equipment 1D data includes specifications, size, manufacturer, spare parts, maintenance record, repair record, etc.; equipment 2D drawings includes equipment structure drawings, installation drawings, photographs, etc.; and equipment 3D information includes equipment 3D visualized model, sound, video, etc.

4.3 Equipment Maintenance Process Simulation

Nuclear power station equipment maintenance process simulation is to reappear or preview the real maintenance process in 3D virtual environment, showing the specific operation method of equipment maintenance visually and dynamically, discovering the spatial relationship of equipment, tools and scene in the process of maintenance, reminding the key operation points and risk precautions, so as to standardize and visualize the maintenance process and improve the efficiency and quality of maintenance work. The specific steps are as follows:

1. Decompose the process of equipment maintenance scientifically and reasonably according to the nuclear power station equipment maintenance procedures, process and safety standards;
2. Build the 3D elements which need to be displayed in maintenance process through 3D modeling, including field, equipment and spare parts, tools and appliance, etc.;
3. Build maintenance simulation animation according to decomposed maintenance process and 3D elements by 3D scene editor, including particle and physical effect and animation control parameters;
4. Develop simulation application function through 3D digital engine, IDE integrated development environment, such as equipment assembly, process jump, pause, text, voice, maintenance record, quality control points, etc.

4.4 Equipment Operation Principle Simulation

Nuclear power station equipment operation principle simulation is to reappear the working principles and effects of each equipment in 3D virtual environment, showing the operation process of the equipment visually and dynamically, providing a visualized analysis method for mastering equipment basic structure and principle, analyzing the key parts and reasons which influence equipment efficiency during operation process.

Nuclear power station equipment operation principle simulation can show the principle of nuclear power station, structure and working mode of the main system and operation principle of single equipment; and can also achieve 3D simulation of nuclear power station equipment operation principle in different conditions, by mapping, texture, particle effect, physical effect, 3D animation and other methods.

4.5 3D Display of Equipment Operation Real-Time Parameters

3D display of nuclear power station equipment operation real-time parameters is to integrate the 3D model and equipment operation real-time data, monitoring and showing dynamically the operation status of equipment, and alarming and 3D positioning for abnormal data in real time.

Firstly develop one-way access interface program according to the interface protocol provided by nuclear power station DCS system and the equipment field bus control system as well as the information security management regulations; then develop function program through real-time data obtained by interface program in 3D IDE integrated development environment.

5 Conclusion

The system is based on virtual reality technology and 3D digital technology applied in nuclear power station equipment technical management, building a integrated application platform for nuclear power station dynamic virtual simulation, visualizing and standardizing the refined equipment technical management and maintenance process, helping saving maintenance cost, improving equipment technical management and risk prevention ability of nuclear power station. The system can also be used for professional training to improve training quality and equipment operating proficiency.

References

1. Hong-Bo Li (2008) Nuclear power plant's key equipment virtual assembly simulation system [J]. China High-tech Enterprises
2. Lin Sun (2008) Nuclear Power Station Maintenance Quality Management [J]. Nuclear Power Engineering 29(4): 36–38
3. Li Zhu (2002) Research on Technology of Building Virtual Roaming System [J]. Xi'an, Northwestern Polytechnical University

HFE Study About Environment Design Indexes for Main Control Room of Nuclear Power Plant

Bo Cheng, Jian-bo Zhang, Shi-bo Mei, Gang Zhang, Yan Wang and Yu Luan

Abstract Main control room (MCR) is an important site of the nuclear power plant (NPP), and the environmental condition in MCR shall be benefit to operators performing their task effectively and comfortably. With the development of HFE (human factor engineering), there is more attention to environment design of MCR. the MCR environment design mainly include noise reduction, illumination, air condition, interior and colour design, earthquake protection, fire protection, radiation protection, missile protection, and so on. At present, the indexes for environment design of domestic and international regulations and standards are inconsistent. None of regulations and standards is comprehensive, detailed and fully adapting to the MCR environment design of NPP, and some indexes are difficult to achieve. This paper compares the indexes of MCR environment of regulations and standards, performs HFE study about the indexes of MCR environment design combining with engineering experience in different NPPs. Finally, this paper proposes a set of environment design indexes applied to MCR of NPP, which will guide the MCR environment design of third generation NPPs.

Keywords NPP · MCR · HFE · Environment design indexes

1 General

The MCR environment of NPP affects the health, performance and human factors effect of operators, even the safety of NPPs. For the MCR design, firstly to complete the basic structure and equipment arrangement of the control room according to HFE theory, then the MCR environment design consider the following indexes: noise reduction, illumination, air conditions, interior and colour design, earthquake protection, fire protection, radiation protection, missile protection, and so on. But now, the indexes of MCR environment design in different regulations and standards

B. Cheng (✉) · J. Zhang · S. Mei
G. Zhang · Y. Wang · Y. Luan
China Nuclear Power Engineering Co., Ltd., Shenzhen 518172, China
e-mail: chengbo@cgnpc.com.cn

Y. Xu (ed.), *Nuclear Power Plants: Innovative Technologies for Instrumentation and Control Systems*, Lecture Notes in Electrical Engineering 400,
DOI 10.1007/978-981-10-3361-2_2

are inconsistent. None of regulations and standards is comprehensive, detailed and fully adapting to the MCR environment design of NPPs, and some indexes are difficult to achieve. So it's necessary to establish a set of comprehensive, consistent and suitable indexes for MCR environment design of NPPs.

2 MCR Environment Design Indexes in Regulations and Standards

This chapter compare and analyse the indexes for MCR environment in regulations and standards to expound the present situation of the environment design requirements of the MCR.

2.1 *Noise Level*

Noise is an important index that affects the environment of MCR in NPP and a major contributor to operator fatigue and effectiveness. Noise reduction is the emphasis and difficulty of MCR environment design. The indexes of MCR noise level in regulations and standards are shown in Table 1.

Table 1 The noise level indexes in the regulations and standards

Regulations and standards	Indexes
IEC 60964-1989 [1]	Maximum ambient noise: 45 dB Background noise should not exceed 45 dB
IEC 60964-2009 [2]	Guidance for environmental specifications under normal conditions is provided in ISO 11064
ISO 11064-6:2005(E) [3]	The ambient noise should not exceed 45 dB $L_{Aeq,T}$ The background level should be in the range 30–35 dB $L_{Aeq,T}$
EUR [4]	MCR shall have an ambient noise level no greater than 50 dB
URD [5]	MCR shall have an ambient noise level no greater than 60 dB (A) Mid-frequency reverberation times should not exceed 0.75 s and should preferably be closer to 0.4 s—dependent on room
Nureg0700 [6]	Background noise levels should not exceed 65 dB (A) The acoustical treatment of the control room should limit reverberation time to 1 s or less
HAF J0055-1995 [7]	Ambient noise should not exceed 45 dB Reverberation (Chinese word is "huisheng" in this standard) time should be limited to below 1 s
GB/T 13630-1992 [8]	Equivalent to the requirements in IEC 60964-1989
EJ/T 638-1992 [9]	Background noise in the control room should not exceed 45 dB (A) Reverberation (Chinese word is "huixiang" in this standard) time should not exceed 1 s
NB/T 20190-2012 [10]	Noise limitation of MCR is 55 dB (A)

According to Table 1, the noise indexes mainly include ambient noise, background noise level and reverberation time, but the indexes in different regulations and standards are inconsistent. The suggesting indexes will be provided in Sect. 4.

2.2 *Illumination*

In normal condition, normal lighting works in MCR, but the possibility of losing normal lighting should be considered, so emergency lighting is as a backup. The indexes of MCR illumination in regulations and standards is shown in Table 2.

Table 2 The illumination indexes in the regulations and standards

Regulations and standards	Indexes
IEC 60964-1989 [1]	Level of illumination: minimum: 200 lx, maximum: 750 lx Uniformity of illumination (ratio): not less than 0.5 Incident illumination to VDU screen: minimum: 50 lx, maximum: 100 lx Minimum emergency illumination: 200 lx
IEC 60964-2009 [2]	Guidance for environmental specifications under normal conditions is provided in ISO 11064
ISO 11064-6:2005(E) [3]	Illuminance levels on work surfaces where paperwork is undertaken should be "maintained" at a level of 200–750 lx with an upper limit of 500 lx where VDUs are used Dimming should be provided with a lower limit of "maintained" 200 lx on the work surface at all times For working areas where mainly paperwork is undertaken, an illumination level of 500 lx should be maintained Electric lighting should achieve a glare index (UGR) of 19 or less for all work positions
EUR [4]	Illumination for the lighting areas (e.g. Operator workstations, auxiliary and supervisor areas) may be in the range of 50–250 lx More detailed guidance is given in IEC 60964: Design for control rooms of nuclear power plants
URD [5]	Control stations shall be provided with lighting which can be adjusted by the operators to provide uniform illumination in the range of 10–50 foot-candles The emergency lighting system shall provide a minimum illumination level of 10 foot candles Note: 1 footcandle = 10.761 lx
Nureg0700 [6]	Nominal illumination levels for various tasks and work areas indicate in Table 1 of Chapter "A Study about Software Development QC and QA of the Digital RPS in Nuclear Power Plant" are 50 or 100 foot-candles, reading on VDU: 10 footcandles Emergency operating lighting: 10 footcandles

(continued)

Table 2 (continued)

Regulations and standards	Indexes
HAF J0055-1995 [7]	Average illumination is 100–500 lx, adjustable Uniformity of illumination is not less than 0.7 Incident illumination (Chinese word is "shigu zhaodu" in this standard) to VDU screen is 50–100 lx Emergency illumination is not less than 200 lx
GB/T 13630-1992 [8]	Equivalent to the requirements in IEC 60964-1989
EJ/T 638-1992 [9]	The illumination requirements are from 200 to 1000 lx distinguishing between different areas, and please see the original text for details Emergency illumination is not less than 200 lx and can work 8 h at least

According to Table 2, the illumination indexes mainly include nominal illumination level, emergency illumination level, uniformity of illumination and UGR, but the indexes in different regulations and standards are inconsistent. The suggesting indexes will be provided in Sect. 4.

2.3 Air Conditions

The MCR of NPP should have appropriate temperature, humidity and air flow organization. The indexes of MCR air condition in regulations and standards are shown in Table 3.

Table 3 The air condition indexes in the regulations and standards

Regulations and standards	Indexes
IEC 60964-1989 [1]	No requirements
IEC 60964-2009 [2]	Guidance for environmental specifications under normal conditions is provided in ISO 11064
ISO 11064-6:2005(E) [3]	For sedentary activity during winter conditions: the operative temperature should be between 20 and 24 °C (i.e. 22 ± 2 °C) The mean air velocity should be less than 0.15 m/s The relative humidity should be between 30 and 70% For sedentary activity during summer conditions: the operative temperature should be between 23 and 26 °C (i.e. 24.5 ± 1.5 °C) The mean air velocity should be less than 0.15 m/s The relative humidity should be between 30 and 70%
EUR [4]	Applicable values for the temperature of the HMI rooms are in the range of 18–25 °C with a humidity rate of 40–60%
URD [5]	73–78 °F; 35–50% relative humidity

(continued)

Table 3 (continued)

<table>
<tr><th>Regulations and standards</th><th>Indexes</th></tr>
<tr><td></td><td>Within a temperature range of 73–78 °F with a 1 h maximum of 85 °F and a relative humidity range of 25–60% maximum
For a loss of all ac power event, provisions shall be made to limit the average temperature rise to 15 °F maximum at 72 h into the event
Note: conversion formula of temperature: F = 9C/5 + 32</td></tr>
<tr><td>Nureg0700 [6]</td><td>The climate control system should maintain temperatures of 68–75 °F in winter and 73–79 °F in summer and relative humidity levels between 30 and 60%
The ventilation system should be capable of introducing fresh air into the control room at a rate of at least 20 cubic feet per minute per occupant</td></tr>
<tr><td>HAF J0055-1995 [7]</td><td>Temperature: 18–25 °C;
Relative humidity: 20–60%
The mean air velocity through the body of main area should be less than 15 m/min. Introducing fresh air into the control room is at a rate of at least 0.43 m^3 per minute per occupant</td></tr>
<tr><td>GB/T 13630-1992 [8]</td><td>No requirements</td></tr>
<tr><td>EJ/T 638-1992 [9]</td><td>Temperature: 20–25 °C
Relative humidity: 30–60%
The mean air velocity through the body of main area should be less than 0.23 m/s
In normal situation, introducing fresh air into the control room is at a rate of at least 0.43 m^3 per minute per occupant
Temperature requirements in extreme environment
<table><tr><td>Duration time of work: h</td><td>1</td><td>2</td><td>4</td><td>12</td></tr><tr><td>Extreme maximum temperature: ℃</td><td>38</td><td>35</td><td>32</td><td>29</td></tr><tr><td>Extreme minimum temperature: ℃</td><td>0.1</td><td>13</td><td>15</td><td>17</td></tr></table></td></tr>
</table>

According to Table 3, the air condition indexes mainly include temperature, humidity, the mean air velocity and introducing fresh air requirements, even the temperature requirements in extreme environment. But the indexes in different regulations and standards are inconsistent. The suggesting indexes will be provided in Sect. 4.

2.4 *The Interior and Color Design*

The technical indexes of MCR decoration material of NPPs are an important index in the interior design. In addition to considering the basic requirements of human health, fire, etc., but also need to be combined with the principles of human engineering and effectiveness.

The reflection coefficient of a variety of colours should be considered in colour design, and the choice of main tone is also consider the user's needs to provide a pleasant working environment and a calming backdrop. The requirements of interior and colour design in regulations and standards is not too much and shown in Table 4.

Table 4 The interior and color design indexes in regulations and standards

<table>
<tr><th>Regulations and standards</th><th>Indexes</th></tr>
<tr><td>IEC 60964-1989 [1]</td><td>No requirements</td></tr>
<tr><td>IEC 60964-2009 [2]</td><td>No requirements</td></tr>
<tr><td>ISO 11064-6:2005(E) [3]</td><td>In selecting materials and finishes for the control areas the following should be considered
(a) The reflectance value of the floor finishes should be between 0.2 and 0.3
(b) Wall finishes should have a surface reflectance of between 0.50 and 0.60. The surface reflectance value should not fall below 0.50, as values below this can increase the contrast between the ceiling and walls, contribute to a gloomy environment, and increase electric light power consumption
(c) The glazing bars and solid areas of the partitions should have a similar reflectance value (0.5–0.6) to the periphery walls
(d) Where indirect lighting systems are used, ceilings should be white, should be of matt finish and should have a minimum surface reflectance of 0.8</td></tr>
<tr><td>EUR [4]</td><td>Typically the floor covering material and the flooring system provide resistance values from 50 to 100 MΩ measured between two points on the floor spaced by 1 m</td></tr>
<tr><td>URD [5]</td><td>General requirements of material storage and surface finishes but no indexes</td></tr>
<tr><td>Nureg0700 [6]</td><td>Reflectances
<table>
<tr><th>Surface</th><th>Preferred (%)</th><th>Permissible (%)</th></tr>
<tr><td>Ceiling</td><td>80</td><td>60–95</td></tr>
<tr><td>Upper wall</td><td>50</td><td>40–60</td></tr>
<tr><td>Lower wall</td><td>15–20</td><td></td></tr>
<tr><td>Instruments/displays</td><td>80–100</td><td></td></tr>
<tr><td>Cabinets/consoles</td><td>20–40</td><td></td></tr>
<tr><td>Floor</td><td>30</td><td>15–30</td></tr>
<tr><td>Furniture</td><td>35</td><td>25–45</td></tr>
</table>
</td></tr>
<tr><td>HAF J0055-1995 [7]</td><td>No requirements</td></tr>
<tr><td>GB/T 13630-1992 [8]</td><td>No requirements</td></tr>
<tr><td>EJ/T 638-1992 [9]</td><td>Equivalent to the requirements in Nureg0700</td></tr>
</table>

According to Table 4, the interior and colour design indexes mainly include reflectance which can't satisfy the design requirements fully. The detailed suggestions will be provided in Sect. 4.

2.5 Other Requirements

In addition, there are some other requirements for MCR environment design in IEC 60964-1989, IEC 60964-2009, ISO 11064-6:2005(E), EUR, URD, Nureg0700, GB/T 13630-1992, EJ/T 638-1992, e.g. fire protection, radiation protection, missile protection, earthquake protection, prevention of hostile acts, and so on. These requirements are the general or basic requirements for the MCR environment design, and no indexes requirements generally (except the indexes requirements of radiation protection in EJ/T 638-1992). But such general or basic requirements should be complied with. For example, the requirements in EJ/T 638-1992 is following: For discontinuous running equipment in control room, such as cable tray and catheter, ceiling, lamps and lanterns, their fault should not injure operators and damage the safety system function during or after S2 earthquake [9]. So these requirements will not be analysed in this paper.

3 Environment Design Indexes in Different NPPs

This chapter analyse the actual or design indexes of noise, illumination, air conditions, interior and colour design of MCR environment design of Ling Ao phase II (LA II), CPR1000 new project, EPR, AP1000 projects. The general or basic requirements without detailed indexes are not analysed here.

3.1 Noise Level

The actual value or design target value of MCR noise of different NPPs is shown in Table 5.

According to the above situation, the 45 dB index requirement of noise level is difficult to achieve.

During the MCR environment design, a questionnaire investigation was performed with six crews to estimate the noise level of MCR [11].

The estimation criteria are:

- Perfect satisfied design (10)
- Good satisfied design (8–10)
- Satisfied design (6–8)
- Bad design (0–6) (Table 6).

Table 5 The actual value or design target value of MCR noise of NPPs

No.	Project	Indexes
1	Actual value of LA II	Less than 52 dB, reverberation time in some frequency is less than 1 s
2	Actual value of CPR1000 new project	Hong Yanhe: less than 51 dB Fang Chenggang: less than 50 dB; reverberation time is less than 1 s
3	Actual value of Tian Wan NPP	55–60 dB; before improvement, the smooth surround wall and ground, and simple perforated plate with big hole on ceiling have low frequency noise absorption function and perform a large reverberation sound field
4	Design target value of EPR basing on EUR	50 dB
5	Design target value of AP1000 basing on Nureg0700	55 dB (A), peak value is not exceeding 65 dB (A). Reverberation time is not exceeding 1 s
6	Actual value of Ling Ao phase I	55 dB, some measuring points exceed 60 dB
7	Actual value of N4	65 dB

Table 6 The estimation for the noise in MCR

Project	Estimation item	Score
LAII	How about the noise in MCR?	7.4
Hong Yanhe	How about the noise in MCR?	7.8

So the background noise level with about 50 dB is reasonable.

For the LAI, the shift team put forward the requirement to reduce the noise level of MCR, c about 60 dB is not ideal.

3.2 *Illumination*

- Actual value of LA II basing on IEC 60964-1989: normal illumination is from 200 to 750 lx distinguishing between different areas and can be adjusted continually; the emergency illumination in main operation area is about 200 lx.
- Actual value of Fang Chenggang basing on IEC 60964-1989: normal illumination is from 200 to 750 lx distinguishing between different areas and can be adjusted continually; optimal illumination in main digitized operation area is 500 lx, the uniformity of illumination is not less than 0.5; Emergency illumination is not less than 200 lx and can work 8 h at least.
- Design target value of EPR mainly basing on IEC 60964-1989: The illumination level shall be 500 lx in the area of the workplaces and in the other areas.

- Design target value of AP1000 basing on Nureg0700: overall illumination can be adjusted from 250 to 500 lx; illumination of reading zone is form 800 to 1000 lx. Emergency illumination is not less than 100 lx in all areas.

The optimal illumination in main digitized operation area is 350 lx of Hong Yanhe MCR. After commercial operation, a questionnaire investigation was performed that 41.7% operators thinks that such illumination is a little dark. Optimal illumination in main digitized operation area is improved to 500 lx of Fang Chenggang, a questionnaire investigation was also performed that 26.4% operators are no comments because the MCR is not running, but 73.6% operators thinks that such illumination is ideal according to simulator.

3.3 Air Condition

- Actual value of LAII: satisfy the indexes of HAF J0055 in Table 3. But a questionnaire investigation shows that 18 °C is a little cold.
- Actual value of CPR1000: satisfy the indexes of HAF J0055 in Table 3.
- Design target value of EPR mainly basing on EUR: The temperature is between 18 and 22 ± 2 °C max. 24 °C and the relative humidity between 40 and 60% and in accidental conditions between 10 and 30 °C for 24 h.
- Design target value of AP1000: design basing on the indexes of Nureg0700 in Table 3.

3.4 Interior and Colour Design

At the beginning of design of LAII, a questionnaire investigation was performed and shown that the interior and colour design of MCR environment is important. Please see Figs. 1 and 2 and Tables 7, 8 and 9.

The choice of main tone of color is considering the user's opinion. A survey result shows that the blue color image perceived by the operators was classified into fresh, cheerful, and comfortable. So the blue is chosen as the main color of LAII.

For the interior design, it's important to improve working efficiency and make operators feel more comfortable. For the technical indexes of the decoration materials, besides the requirement in Table 4, the functionality should be considered, e.g. the sound absorption index. In Fang Chenggang project, the technical indexes of the decoration materials are optimized entirety. The technical factors of MCR floor is as following table that solves the wear, reflection, antiskid and other issues of ground effectively.

The indexes of above factors are provided according to the standards and engineering experience in NPPs.

AP1000: design basing on the requirements of Nureg0700 in Table 4.

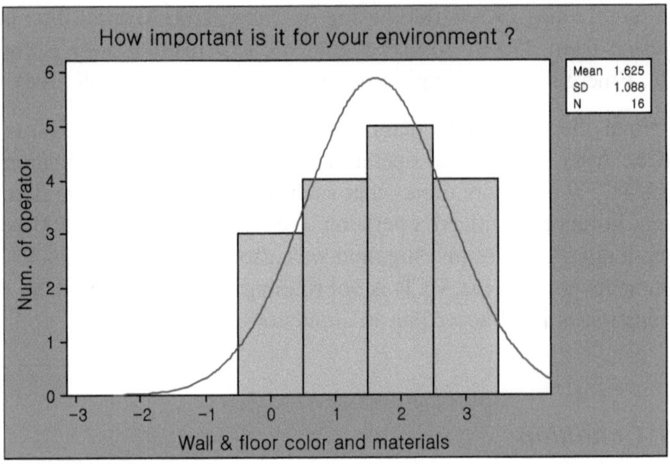

Fig. 1 Questionnaire investigations on wall and floor color and materials

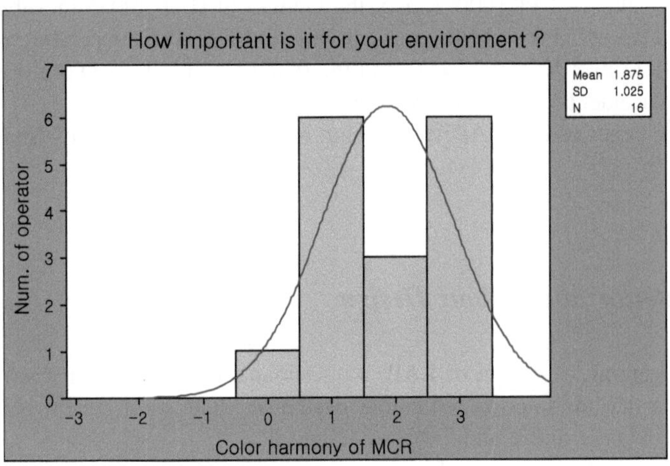

Fig. 2 Questionnaire investigations on color harmony

Table 7 Statistical result of Fig. 1

No.	Number of operator	Mean	*P*-value (0.05)	Result
1	16	1.63	0.00	Important

Table 8 Statistical result of Fig. 2

No.	Number of operator	Mean	*P*-value (0.05)	Result
2	16	1.88	0.00	Important

Table 9 Technical factors of floor material of Fang Chenggang project

No.	Technical factor
1	Specification
2	Color
3	Impact strength
4	Density
5	Water absorption rate
6	Combustion performance
7	Wearing strength
8	Moh's hardness
9	Compressive strength
10	Reflectance value
11	Glossiness
12	Meet the requirements of class A decoration materials standards in GB6566-2010 (limits of radionuclides in building materials)
13	Friction index (wet)

4 Analysis and Project Experience of Environment Design Indexes

Basing on the analysis in Sects. 2 and 3 and the research achievement of NPPs, this chapter put forward that the suggested indexes of noise level, illumination, air condition, interior and colour design of MCR environment.

4.1 Noise Level

Background noise level and reverberation time are suggested to consider for MCR noise level of NPPs. The detailed indexes are as following:

- Background noise in MCR should not exceed 50 dB (A) according to the project experience.
- Mid-frequency reverberation times should not exceed 1 s according to the project experience.

In Table 1, the description of "huisheng" in HAF J0055 and "huixiang" in EJ/T 638-92 is not accurate; the correct interpretation is "hunxiang" originating from word "reverberation". This should be corrected when the regulations and standards are updated.

4.2 Illumination

The MCR should be provided with several lighting areas, which can be manually adjusted to provide illumination suitable for the Operators to perform their tasks.

Suggesting indexes:

- Level and uniformity of illumination are suggested subjecting to IEC 60964-1989 and optimal illumination in main digitized operation area is 500 lx.
- Level and working time of emergency illumination is suggested subjecting to EJ/T 638-1992.
- Uniformity of illumination (ratio) is not less than 0.5 subjecting to IEC 60964-1989.
- For direct light, electric lighting should achieve a glare index (UGR) of 19 or less for all work positions subjecting to ISO 11064-6:2005(E) and project experience.

In Table 2, the description of "shigu zhaodu" in GB13630 and HAF J0055 is not accurate; the correct interpretation is "rushe zhaodu" originating from word "incident illumination". This should be corrected when the regulations and standards are updated.

4.3 Air Condition

The different index of temperature for winter and summer is more reasonable. Suggesting indexes are as following:

- The temperature, humidity and mean air velocity of normal air condition is suggested subjecting to ISO 11064-6:2005(E).
- The rate of introducing fresh air is suggested subjecting to HAF J0055-1995.
- Temperature requirements in extreme environment subjects to EJ/T 638-1992.

4.4 The Interior and Color Design

For the interior design, suggest considering the surface reflectance basing on Table 4 and actual requirements, and specify detailed technical indexes of the decoration materials according to the functionality. The author suggests determining the main tone of colour with the users together to provide a pleasant working environment and a calming backdrop.

4.5 *Others*

The requirements in Sect. 2.5 are general or basic requirements for MCR environment design and should be complied with.

5 Conclusion

The MCR of NPP is an important work place for the operation staff. A good environment of MCR can improve the work efficiency of MCR staff and the safety of NPP. This paper put forward a set of entire, specific and strongly adaptive indexes for MCR environment design. This research achievement lays a solid foundation for the MCR environment design of the third generation NPPs.

References

1. IEC 60964-1989. Design for control rooms of nuclear power plants [S].
2. IEC 60964-2009. Nuclear power plants-Control rooms-Design [S].
3. ISO 11064-6. Ergonomic design of control centres-Part 6: Environmental requirements for control centres [S].
4. EUR 2.10 revision D. Instrumentation & control and human-machine interface [S].
5. URD revision 8 volume III chapter 1. Overall requirements [S].
6. Nureg0700 Rev.2. Human-System Interface Design Review Guidelines [S].
7. HAF.J0055-1995. Engineering principles for control room design of nuclear power plant [S].
8. GB/T 13630-1992. The design of control room of nuclear power plant [S].
9. EJ/T 638-1992. Design criteria for control room complex of nuclear power plant [S].
10. NB/T 20190-2012. Noise control for production buildings of nuclear power plants [S].
11. Pengfei Gu, etc. Study about Ceiling Design for Main Control Room of NPP with HFE. Engineering Village (2014) Energy Materials Conference Proceedings.

References

A Study of Implementation V&V Activities for Safety Software in the Nuclear Power Plant

Hui-hui Liang, Peng-fei Gu, Jian-zhong Tang and Wei-hua Chen

Abstract In order to improve the safety and reliability of the safety related software for the nuclear power plant, the verification and validation (V&V) is necessary for the software in the whole development life cycle based on nuclear laws and principles requirement. Nuclear Safety software implementation phase means using programming language to perform system requirement specifications. Implementation V&V activities based software design documentation to verification and validation the correctness, accuracy and completeness of the source code. This paper considered the regulation and law requirements and given the test types of implementation V&V activities. The test process of implementation V&V activities for safety software in nuclear power plant has been given. The paper also proposed the key points of the test case and the principles that the test person need to follow.

Keywords Safety software · Implementation V&V activities · Test type · Execute

1 Introduction

The safety and reliability of nuclear safety application software will directly affect the entire nuclear power plant safety and economy. To ensure the quality level and improve nuclear safety level of reliability, nuclear safety application software need to perform independent software verification and validation (V&V) work within the entire development life cycle. Nuclear power plant safety applications software V&V work is the key point to self-developed nuclear safety-class system.

H. Liang (✉) · P. Gu · J. Tang · W. Chen
State Key Laboratory of Nuclear Power Safety Monitoring Technology and Equipment, China Nuclear Power Design CO., LTD, Shenzhen, China
e-mail: lianghuihui@cgnpc.com.cn

Y. Xu (ed.), *Nuclear Power Plants: Innovative Technologies for Instrumentation and Control Systems*, Lecture Notes in Electrical Engineering 400,
DOI 10.1007/978-981-10-3361-2_3

According to IEEE 1012 [1], software V&V activities include concept V&V, requirement V&V, design V&V, implementation V&V and integration V&V and so on. The software implementation phase means using programming language to perform system requirement specifications. The implementation phase will generate the source code that the computer can read. Implementation V&V activities are based on software design documentation to verification and validation the source code is correctness, accuracy and completeness.

The V&V standards and regulations are developed by the international organizations, such as IEEE [1, 2], NRC [3] and IEC [4]. The domestic [5–7] is also to compile the standards and guidance documents. The standards and regulations only give the software V&V activities and performance requirements. They don't carry out the system implementation process and what test type need to be executed for the software implementation phase V&V.

The paper is organized as follows. The first part introduced the research background. The second part collates the domestic and international standards and regulations for the safety application software V&V test. The second part gives the test types that need to be performed in the safety application software implementation V&V. The third part presents the implementation phase V&V test process and methods which considers the test type. The fourth part is the conclusion.

2 Software Test Requirements in Regulations and Standards

In order to achieve the better results of nuclear power plant safety software V&V activities, the nuclear advanced countries build the nuclear power plant V&V regulations and standards. The International Atomic Energy Agency (IAEA) and International Electro technical Commission (IEC) promulgate the nuclear power plant software V&V standards and regulations. IEC 60880 [4] gives the rules of design and documentation for the category A functions software. It includes specification, design, verification and validation, etc. IEEE 1012 [1] is the verification and validation standard for system and software. It regulates the content, scope, method and demand of the software implementation V&V activities. It also gives the minimum task set. IEEE 1028 [8] describes the guidance on how to conduct software reviews, inspections, technical and management review. It provides the reference for the implementation V&V activities. IEEE 829 [9] gives the purpose, content, formal and key elements of software test. The NRC R.G.1.170 [10] accepts IEEE 829. The software unit test requirements and method is put forward by IEEE 1008 [11] and R.G.1.171 [12]. The domestic standards and regulations are mostly transformed by IEC. HAD 102/16 [5] provides the verification and validation plan.

Based on NB/T 20054-2011 [7], the nuclear plant A category function software should be divided into module when the program is designed. So the implementation V&V activities for safety-class software need to test the module units before component test.

3 Test Types for Implementation V&V Activities

The application software implementation V&V activities include the module unit, function, interface and component test. The unit test is to verify the software units satisfy the software specifications and coding standards. Currently the test types for high security application include dozens. The paper combines the nuclear safety software its own characteristics and standard and regulation requirements that gives the test types in the software implementation V&V activities.

The standard and regulation requirements provide the following demands for the software and documents:

1. The software satisfies the software specification and it can be verified. The code can also be read, understood and has adequate comments.
2. The code should be simple. The recursive structure, code compression and optimization need to be avoided.
3. The data structure and name should be consistent in the whole system. It includes data type, scope, accuracy and so on.
4. In order to regulate and guide the program design, the approved encoding rules are requisite.
5. The software should be testability that means the every executable code can be interviewed.
6. The document should include software design specification and logical requirements. It also needs to contain pre-condition and post-condition for every module.
7. The input and output variable, the effective scope and interface should be described in the document.
8. The document and software should give the methods of abnormal situation. The reliability and safety measures are also needed to consider.

Due to the aforementioned conditions, the nuclear safety application software implementation V&V need to performed the minimum set of the test type as shown in Table 1. From Table 1, we can see that the nuclear safety application software implementation V&V need to execute at least thirteen test types.

Table 1 Test types of implementation V&V activities for safety application software in the nuclear power plant

No.	Test type	Requirement specification
1	Document review	Do completeness, consistency, accuracy and correctness review for requirement documents
2	Code review	Review the code consistent with requirement documents and standards and code structure design is reasonable and code readability
3	Static analysis	It includes data flow analysis, control flow analysis, and information flow analysis
4	Code walkthrough	Code walkthrough is a way of checking the code design flaw. Code walkthrough executes the test case by human brain to verify the code to run properly and meet the functional requirements
5	Logical test	Logical test is to verify the software logical structure is correct. It is able to achieve all branch and statement coverage and make sure that the code does not exist any infinite loops or redundant branches or statements
6	Function test	Functional test needs to verify every functional requirement. Functional test needs to verify the normal and abnormal situation for every module
7	Interface test	Interface test should perform the conformance between the software interface data types and structures. Input and output interfaces for each module should be executed the abnormal and normal test
8	Reliability test	The specific test need to be performed for the situation which may change the operation mode. It includes the boundary and environmental test. The human negligence design which will affect the software reliability that should be avoided, such as division by zero
9	Regulatory compliance test	The compliance between the application software design and the development standards should be executed. Standards compliance evaluation criteria should be established
10	Compatibility test	Nuclear safety application software compatibility test mainly refers to the new version application software to retain the old version function. In this case, it needs to verify the compatibility between the two versions, the potential risks and incompatibilities
11	Performance test	Nuclear safety application software implementation V&V performance test is mainly for the response time, accuracy of the output data and other critical functions
12	Boundary test	Boundary test is the running state test when the software in the boundary and endpoint. It includes state transition boundary, functional boundary and properties boundary
13	Data processing test	Data processing test includes the conversion of test data, test data acquisition and the capability of removing the bad data

4 Implementation V&V Activities

Nuclear Safety software implementation V&V test technology and methods can be divided into static test and dynamic test. The nuclear safety software implementation V&V test can be divided into document review, code walkthrough and code analysis, unit test and component test.

1. Document review

It is mainly to review the software module unit document in the correctness, completeness, accuracy, consistency. The tester need to verification and validation the demands of software module unit documentation are traceable.

2. Code walkthrough and code analysis

Code walkthrough and code analysis is static test. Static test is not running the program. It executes the program and document analysis and examination. The analysis can use the automation tools which have been reviewed by the project in order to improve the efficiency and quality assurance.

Code walkthrough and code analysis focus on the following problems:

- The document describes inconsistent;
- Code is inconsistent with the requirements document;
- The compliance between code implementation process and the standard;
- Code structure problem;
- Code protection mechanism.

3. Unit test and component test

Unit test and component test is dynamic test. Dynamic test runs the program and achieves the software errors and defects by executing a set of instance data. Software coverage index is a common way to assess the adequacy of the dynamic test. The design of nuclear safety application software test case should consider statement coverage, branch coverage and MC/DC coverage.

In order to find as many as the program design flaws, perfect test case is essential. Table 2 is the compared result which is the different team executes the same project. The project is to test the function for the ordering system. Team A fund thirty one problems by implementing five hundred two test cases. But team B just executed three hundred forty three test cases and fund thirty three problems. So the reasonable and effective case will cover various types of test requirements and coverage metrics, reduce test effort and efficiency work.

Table 2 Compared team A with team B in the same project

Number	Team A	Team B
Test cases	502	343
Problems	31	33

Nuclear Safety software implementation phase V&V test case design needs to be completed after the software requirements specification has been finished. The following key factors should consider when the nuclear safety software implementation V&V design the test case.

- The normal and abnormal value design;
- Each test case needs as many contain multiple test types;
- It need to give a reasonable and feasible evaluation criteria for each test case;
- A test case needs to consider the effective equivalence class;
- The test cases need to cover every requirement;
- The test cases preferably prepared by the engineers which have the relevant project experience.

Unit test and component test include the following common problems.

- Functional description is inconsistent with the requirement;
- Data processing exists risk, such as data accuracy;
- Abnormal value or boundaries don't give the protection mechanisms;
- Interface inconsistencies.

5 Quality Assurance

Nuclear Safety software implementation V&V need to write the test plan and instruction. The test plan includes personnel, scope, purpose, schedule, exception handling quality assurance programs. Test instruction used to guide and regulate the implementation of test cases. The tester should ensure completely and correctly execute the test cases. Tester should comply with the following requirements.

- Testers need to organize test procedures, test plan and test management training files before performing the test;
- Test cases need to be completely and accurately execute in step;
- Testers need to record the real test environment and test results to ensure test repeatability;
- Adding or changing the test case need to be confirmed by the technical director.

6 Application

Above implementation V&V activities for safety-class application in the nuclear power plant had been applied to the CPR1000 project. The project involved nearly two hundred thousand code lines and included four versions. The code lines existed small changes because of the software modules increasing and revising. Problem statistics shows in Figs. 1, 2, 3 and 4. The problems of four versions software were one-hundred-three, sixty-six, twenty-six and nineteen as show in Fig. 5.

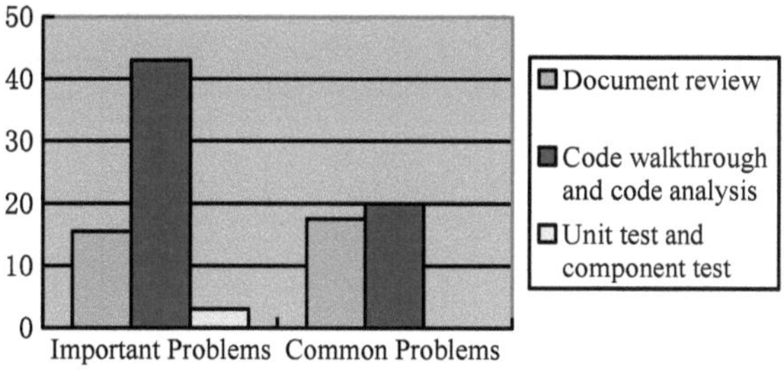

Fig. 1 The first vision software problem statistics

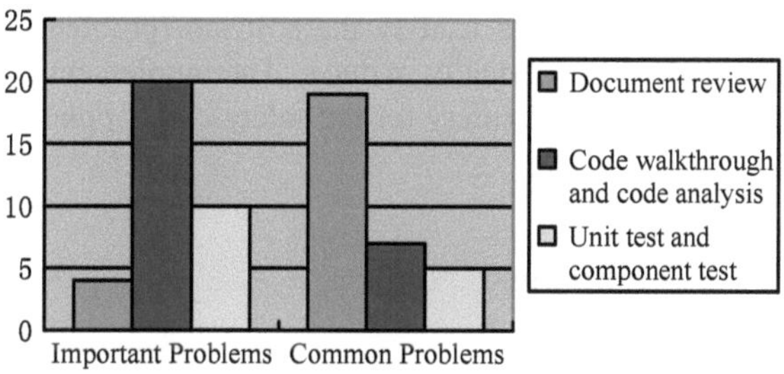

Fig. 2 The second vision software problem statistics

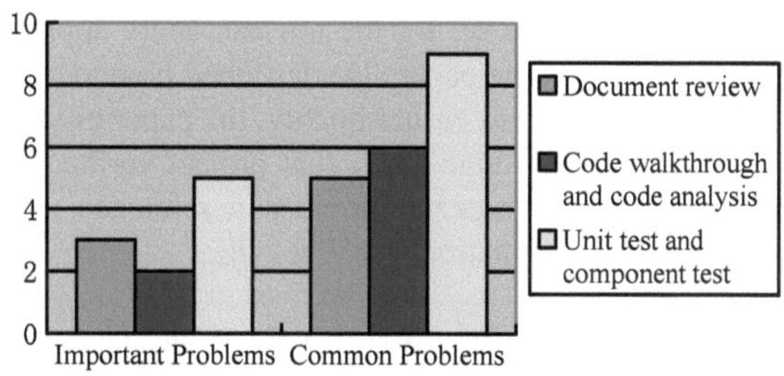

Fig. 3 The third vision software problem statistics

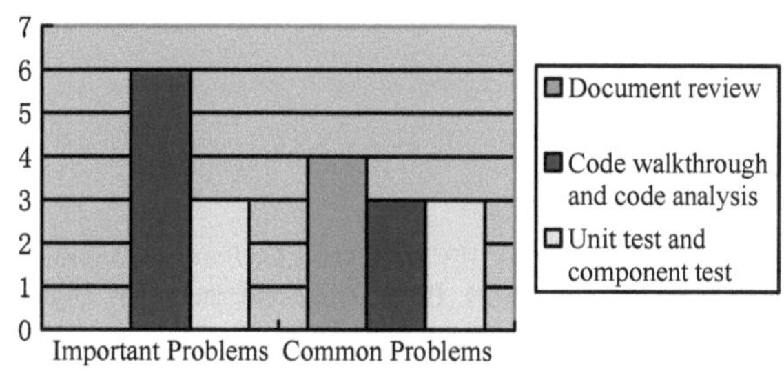

Fig. 4 The fourth vision software problem statistics

From Fig. 1, the problems were focus on software implementation. The test team found sixty-seven problems through code walkthrough and code analysis. The main problems were the process of the documentation requirements transformed into software. Through the three visions testing, the third and fourth visions software problems concentrated on the documentation requirements. That means the

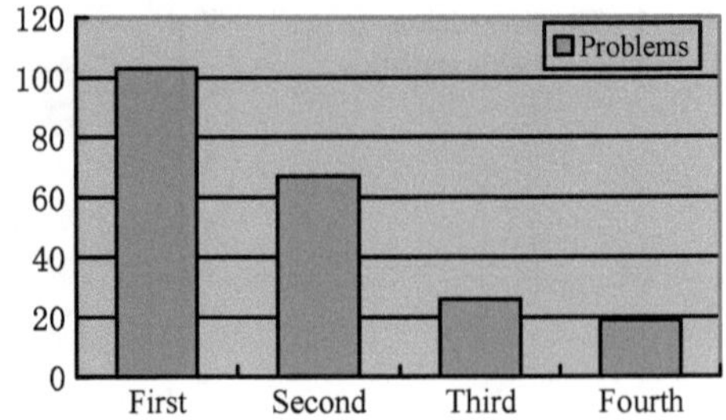

Fig. 5 Problem statistics of four visions

imperfect demands lead to the software problems. As shown in the Fig. 5, the problems are gradually reduce. The implementation V&V activities improved the reliability and safety for the safety-class application in the nuclear power plant.

7 Conclusion

This article had analyzed and discussed the test types for the nuclear safety applications software implementation V&V activities. The paper proposed the minimum test type set for the nuclear safety application software implementation V&V activities. The paper also described how to execute the implementation V&V activities. In order to ensure quality, the paper given the key factor what the person need to obey. Through the actual project verified the effectiveness of implementation V&V activities. It provided a reference and guidance for nuclear safety software implementation V&V activities.

Acknowledgements This project was financially supported with funds which provided by a reliability research of reactor protection system which based on digital microprocessor and electrical equipment (national science and technology major project). The award number is 2014ZX06004002-004.

References

1. IEEE 1012 (2004). IEEE Standard for Software Verification and Validation.
2. IEEE 7-4.3.2 (2010). IEEE Standard Criteria for Digital Computers in Safety Systems of Nuclear Power Generating Stations.
3. R.G.1.168 (2013). Verification, Validation, Reviews and Audits for Digital Computer Software Used in safety Systems of Nuclear Power Plant.
4. IEC 60880 (2006). Nuclear Power Plants Instrumentation and Control Systems Important to Safety – Software Aspects for Computer-Based Systems Performing Category A Functions.
5. HAD 102/16 (2004). Systems Important to Safety Based on Computer of Nuclear Power Plants.
6. HAF102 (2004). Regulate for Nuclear Power Plant Design Safety.

7. NB/T 20054 (2011). Nuclear Power Plants Instrumentation and Control System Important to Safety Software Aspects for Computer based System Performing Category A Functions.
8. IEEE 1028 (1997). IEEE Standard for Software Reviews.
9. IEEE 829 (1998). IEEE Standard for Software Test Documentation.
10. R.G.1.170 (1997). Software Test Documentation for Digital Computer Software used in Safety Systems of Nuclear Power Plants.
11. IEEE 1008 (1987). IEEE Standard for Software Unit Test.
12. R.G.1.171 (1997). Software Unit Test for Digital Computer Software Used in Safety Systems of Nuclear Power Plants.

Analysis of CPU Redundant Configuration for the Safety DCS of NPP

You-ran Li, Wei Sun, Liang Zhou and Long-qiang Zhang

Abstract Safety digital control system (DCS) has been used in nuclear power plant (NPP). DCS provides different control, monitoring and operating means for nuclear power plant (NPP). It is necessary to consider the redundant configuration for CPU, because CPU is one kind of the important equipment of safety DCS which need to accord with the single failure criterion and reliability requirement. Based on the project technical plan, the paper analyzes the different redundant configuration for CPU of safety DCS system and gives some thoughts on engineering practice. It will be benefit to the following project practice.

Keywords Nuclear power plant (NPP) · Safety DCS · CPU · Single-failure · Redundant configuration

1 Introduction

In recent years DCS technology has be used in all the newly-built nuclear power plant (NPP). It is the important system for the safety operation of the NPP, because it provides an advanced control, monitoring and managing means for the technics and equipment control of nuclear power plant. The CPU is the vital part of the DCS and it need to accord with the single failure criterion and reliability requirement. So redundancy configuration shall be employed for some important CPU structure. It can avoid the failure of the system and the failure of safety functions when to meet the Single-Failure. It help to improve the reliability of DCS functions. Based on the project technical plan, the paper analyzes the different CPU redundant configuration for safety DCS system and gives some beneficial reference for the following DCS design and project practice.

Y. Li (✉) · W. Sun · L. Zhou · L. Zhang
State Key Laboratory of Nuclear Power Safety Monitoring Technology and Equipment, ShenZhen China Nuclear Power Design Co. Ltd., Shenzhen, China
e-mail: liyouran@cgnpc.com.cn

Y. Xu (ed.), *Nuclear Power Plants: Innovative Technologies for Instrumentation and Control Systems*, Lecture Notes in Electrical Engineering 400,
DOI 10.1007/978-981-10-3361-2_4

2 Redundant Configuration Analysis

2.1 The Regulations and Standards

According to the single failure criterion requirements of the regulations, it needs consider the redundant configuration in the design of safety DCS system. The following requirements about the protection function design which from the HAF102 and HAD102 will point out the importance of the redundant configuration.

HAF102 requires that "The means for shutting down the reactor shall consist of at least two different systems to provide diversity. At least one of the two systems shall be, on its own, capable of quickly rendering the nuclear reactor subcritical by an adequate margin from operational states and in design basis accidents, on the assumption of a single failure" [4]. In order to verify whether it meets the single failure criterion, it needs to consider the failure mode and affect for all the equipment and modules of the safety DCS. The reactor trip control system and the safety-related engineered safety feature actuation system shall follow this requirement.

HAD102/10 requires that "Diversity is applied to redundant systems or components that perform the same safety function by incorporating different attributes into the systems or components. The reliability of systems can be improved by maintaining the features for independence in design" [5]. HAD102/14 requires that "Redundancy is commonly used in I&C systems important to safety to achieve system reliability goals and/or conformity with the single failure criterion. For redundancy to be fully effective there should be independence. Higher reliability of the safety-related I&C system is achieved by the use of redundancy or diversity with multiple channels or modules" [6]. Besides the single failure criterion, independence and consistency criteria still are considered for the safety-related I&C system [2].

2.2 Requirements for Design

For the requirements from the regulations and design documents, it needs to meet the Single-Failure Criterion in safety DCS systems performing safety functions and certain safety-related functions [1, 3]:

1. This implies a necessity for redundant design for the protection system. If there is a failure on one of the equipment or channel, the protection function still is performed.
2. This implies a necessity for redundant design for the engineered safety feature actuation system. If there is a failure on one of the equipment or channel, the engineered safety features actuation function still is performed.

3. Redundancy shall be employed in other important I&C system to improve the reliability of safety DCS functions.

Using the PSA modeling analysis for safety DCS, it discovers that the CPU unit is the vital part of the safety DCS, and redundancy configuration should be employed for some important CPU structure. Because it very important for the reliability of safety DCS system.

3 Modes of Redundant Configuration

3.1 Parallel Redundant

For the parallel redundant, the control units which complete redundant each other achieve their respective functions independent, and the realization and output of each unit is independent of the other.

Adopting the parallel redundant mode, the control units in the same group will keep independent and asynchronous operation each other. They receive the same input signals, perform the same logic operation and export the results respectively. Then they will take a vote for the redundant results with the vote logic which is built by another software or hardware.

There should be the versatility of redundant components. And they can meet the requirements of the exchange without changing the function logic code. They can switch each other through simple settings such as the dial switch, switching cards etc. (Fig. 1).

3.2 Control/Standby Redundant

For the control/standby redundant, they have two identical hardware and software and the two control units which are mutual redundancy share the same set of acquisition and output channels. They determine the control-standby through the information exchange, and choose the effective output. In this configuration, the following requirements need to be considered:

1. The requirements of bumpless switchover

Bumpless switchover shall be ensured and the changeover between the control/standby does not allow the signal to jump. The control/standby redundant just is established in the normal operation.

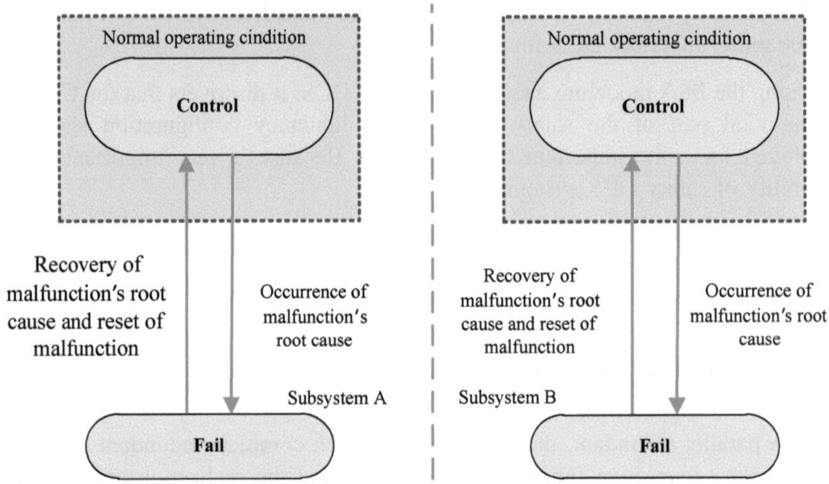

Fig. 1 CPU status transition diagram of parallel redundant system

2. The changeover of the control/standby redundant

The changeover is performed automatically just at the time that the control unit have the detectable fault and the standby unit has no fault. The changeover can be performed manually if necessary.

3. Operation mode

The two subsystems for the control/standby redundant all receive the input date and perform the operation. But only the control subsystems gives the finally output single.

The control/standby states of the system can be identified through certain mechanisms in the normal operation time. It should avoid that all the unit are in the control state.

4. Keep the data synchronization

It needs to consider data synchronization between the two subsystems for the control/standby redundant in each operation cycle.

In addition to the state information and the data synchronization, there should not have any other interaction between the control/standby redundant subsystems. Especially it cannot rely on other subsystem data to complete the protection function of subsystem self (Fig. 2; Table 1).

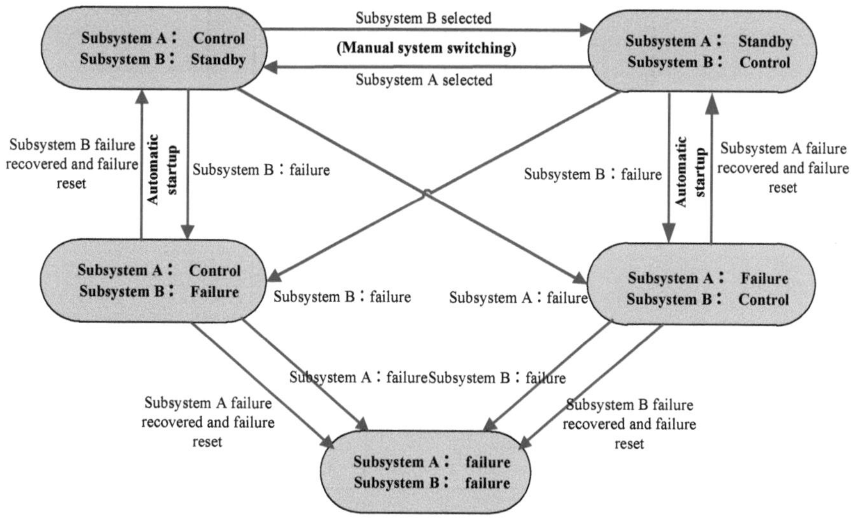

Fig. 2 CPU status transition diagram of control/standby redundant system

Table 1 Comparison between the two modes

Mode	Configuration	Operation	Output
Parallel redundant	Complete redundant each other achieve their respective functions independent	Keep independent and asynchronous operation each other	Output of each subsystem is independent
Control/standby redundant	Have two identical hardware and software and the two control units which are mutual redundancy share the same set of acquisition and output channels	Have the different state at the same time Need to keep the data synchronization and bumpless switchover	Only the control subsystems gives the finally output

4 Analysis of Project Practice

4.1 Parallel Redundant in Project Practice

According to the safety DCS system from one of the CPR1000 nuclear power project, they have the parallel redundant configuration for the CPU of the ALU unit which is using for logic processing. There are two controllers like ALUx1 and ALUy1 for every subgroup with the parallel redundant. Every ALU unit has a CPU card. The finally output signal come from the redundant CPU after a "OR" logic processing (Fig. 3).

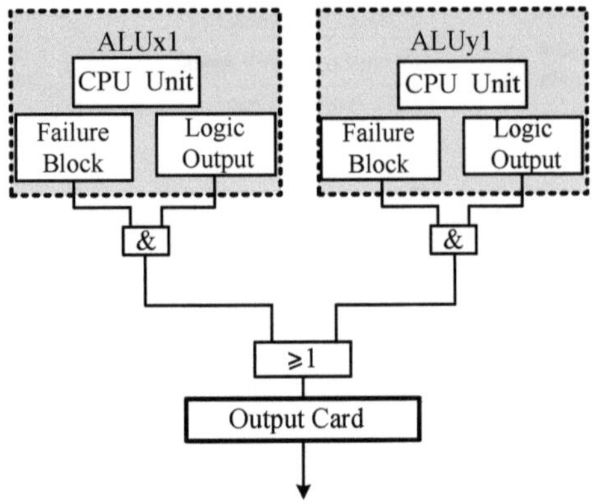

Fig. 3 Equipment configuration diagram of the parallel redundant control cabinet

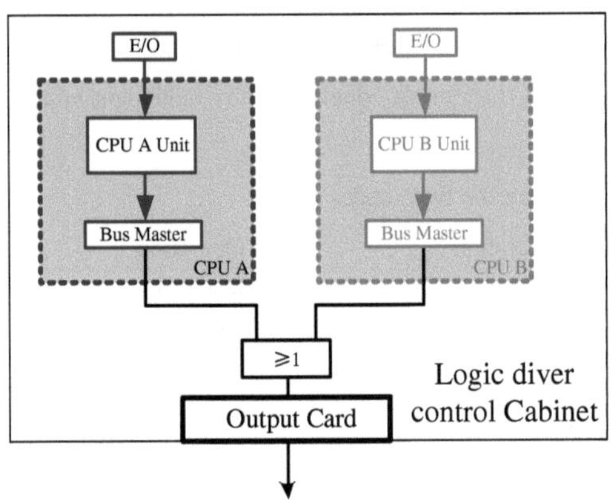

Fig. 4 Equipment configuration diagram of logic diver control cabinet

Another example in project practice likes that: Parallel redundant CPU which comes from two groups receives the same input signal, and gives the result after the operation processing. The finally command signal will be output after a "OR" logic which is made up of hard wired (Fig. 4).

4.2 Control/Standby Redundant in Project Practice

According to the design scheme of safety DCS system from one of the CPR1000 nuclear power plant, they have the control/standby redundant configuration for the CPU of the RPC cabinet which likes the following diagram. The system management card for each CPU can give the operation state monitoring to the Switch Changeover card which will perform the changeover between the control/standby CPU (Fig. 5).

Fig. 5 Equipment configuration diagram of RPC cabinet

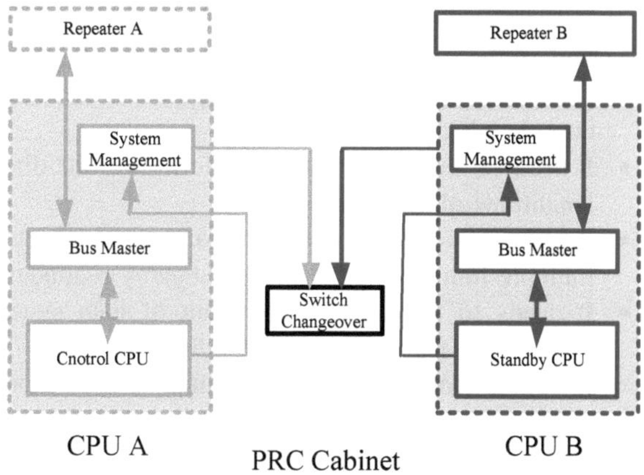

4.3 The Problems from Different Configuration

1. Parallel redundant

For the parallel redundant, it's completely same redundant. One of them is correct and the all unit is effective. It depends on the transition between the control state of and standby state. It needs to consider the double DO problem as the following example:

- Taking an example from a project: In the same subgroup, there are two controllers with the parallel redundant. Every controller has the module which can identify the failure state. When one of controllers is failure, there is a "AND" logic between the failure signal and the real output signal. It can avoid the double DO problem, because the finally output signal is blocked.
- Taking an example from another project: In the same subgroup, there are two controllers with the parallel redundant. When one of controllers is failure and the other is correct, there will be two different output signals from different controllers. Because the failure signal is not blocked. So it would have the double DO problem for the finally output signal.

This problem reminds that the fault recognition of parallel redundant is important for the final correct output. Different types of equipment have the different ways to achieve, and the design should combine with the product characteristics.

2. Control/standby redundant

For the control/standby redundant, it's not a completely same redundant. It depends on the transition between the control state of and standby state. It needs to consider effect of the main and standby switching process from the detectable fault and undetectable fault of the CPU unit. Give a detailed analysis of the main and standby switching with the concrete conditions of switching mode and consider the following questions:

- If it meets the conditions of the changeover just because of some undetectable fault of the control CPU unit, the switch is fail to diagnose the fault and have no changeover. Whether the standby part is deemed invalid and the entire unit is considered failure?
- It needs to know the failure models of all the diagnostic points which meet the conditions of the changeover.
- Whether the more equipment such as switch etc. will bring more the risk of multiple failures for the control/standby redundant?
- It needs to consider the problem of data synchronization between the two subsystems for the control/standby redundant. The changeover shall be performed smoothly and avoid the impact of the system disturbance caused by the changeover.

Generally these problems should been considered in the process of systematic safety analysis. The DCS products supplier needs to provide detailed information about the equipment performance and the fault mode analysis for the safety analysis.

5 Conclusion

In view of the above analysis, there have the diversities for the CPU redundant configuration of safety DCS based on different projects which use the different platforms and different technical schemes. Each redundant configuration has its advantages. In order to choose the suitable CPU redundancy configuration for the safety DCS system, it needs to have the comprehensive evaluation for the different functional requirements and equipment's characteristics for the different project applications.

References

1. General Administration of Quality Supervision Inspection and Quarantine of the People's Republic of China, Standardization Administration of the People's Republic of China (2008). Application of the single failure criterion to safety systems in nuclear power plant [S], GB/T 13626. Beijing
2. General Administration of Quality Supervision Inspection and Quarantine of the People's Republic of China, Standardization Administration of the People's Republic of China (2008). Applicable criteria for digital computers in safety systems of nuclear power plants [S], GB/T 13629. Beijing
3. International Electrotechnical Commission (2011) Nuclear power plants – Instrumentation and control for systems important to safety - General requirements for systems [S], IEC 61513
4. NSSS (2004) Safety of Nuclear Power Plants: Design [Z], HAF 102. Beijing
5. NSSS (2004) Nuclear power plant protection system and related facilities [Z], HAD 102/10. Beijing
6. NSSS (2004) Safety related instrumentation and control system for nuclear power plant [Z], HAD 102/14. Beijing

Research and Improvement of TG Equipment Load Shedding Control Scheme in Nuclear Power Plant

Le-Yuan Bai, Xu-Feng Wang, Gang Yin and Bin Zeng

Abstract Once the emergency diesel generator in operation in CPR1000 nuclear power plant (NPP), the selected equipment reloads in the pre-specified sequence after load shedding, and the important equipment should reload preferentially and successfully. This paper finds that the load shedding command for turbine and generator (TG) equipment is from non-classified turbine control system (NC-TCS) to motor control center (MCC) in a CPR1000 NPP. There is a risk that the more important equipment might fail to reload if TG equipment load shedding fails in case TCS system out of order. This paper adds a new TG load shedding path from class 1E distributed control system (1E-DCS) to MCC, and realizes the hard priority logic of TG load shedding by MCC control loop, increasing the reliability of TG load shedding command and eliminating the influence on other more important equipment. The improvement scheme proposed in this paper enhances the safety of NPP and cuts the cost of nuclear power engineering.

Keywords TG · Emergency diesel generator · Load shedding · Control · Improvement

1 Introduction

To improve the reliability of power supply to key equipment is an important means to ensure the safety of NPP. Normally the power of equipment is supplied by 500 kV offsite power via main transformer and step-down transformer in CPR1000 NPP. When 500 kV off-site power lost, 200 kV off-site auxiliary power will be put into operation via auxiliary transformer. The emergency diesel generator finally takes charge of power supply if 500 and 200 kV off-site power both lost, to ensure the automatic response of related systems, prevent damage of important equipment and ensure the safe shutdown of nuclear power unit [3] (Fig. 1).

L.-Y. Bai (✉) · X.-F. Wang · G. Yin · B. Zeng
China Nuclear Power Engineering Co., Ltd., Shenzhen, China
e-mail: baileyuan@cgnpc.com.cn

Y. Xu (ed.), *Nuclear Power Plants: Innovative Technologies for Instrumentation and Control Systems*, Lecture Notes in Electrical Engineering 400,
DOI 10.1007/978-981-10-3361-2_5

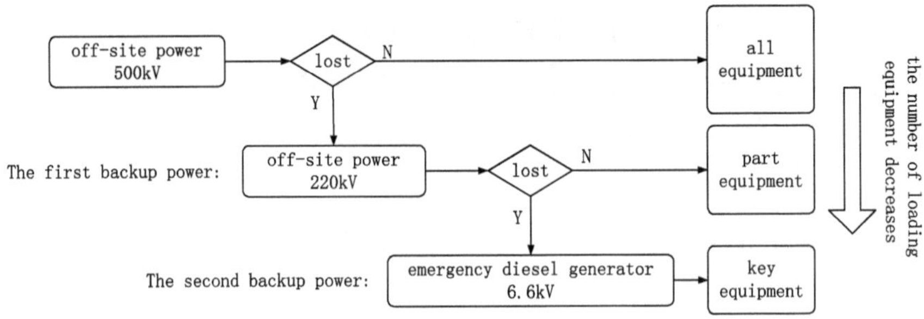

Fig. 1 The power supply of CPR1000 NPP

Table 1 The equipment and time of TG load shedding

No.	Equipment	Quantity	Load shedding time (s)
1	Turbine lube oil pump	1	25
2	Turbine jacking oil pump	2	25
3	Turbine lube oil vapor extractor	1	0
4	Generator seal oil pump	1	40
5	Generator vapor extractor	1	0

Once the power supply is switched to the emergency diesel generator unit, the grid breaker opens and the off-site power of plant is lost. Only certain important equipment starts to reload until the circuit breaker of the emergency diesel generator closes. To make sure of the success of reload, all the pre-selected key equipment has to shed load first and then reload in the pre-specified sequence. The sequence is established according to the importance of the equipment, and the more important the equipment is, the earlier the reload time is. The reload times start from 0 s and increase 5 s delay once. All equipment is forbidden to reload in advance [1, 4].

TG is the most significant equipment of conventional island in CPR1000 NPP. TG important equipment has to be powered by the emergency diesel generator when both 500 kV off-site power and 200 kV off-site auxiliary power fail, so that the turbine can be safely stopped in emergency and accidents can be avoided, such as bush burning, scored cylinder and so on.

2 Analysis of TG Load Shedding Scheme

2.1 TG Load Shedding Scheme

The TG of CPR1000 NPP 1 is a turbine-generator of million-kilowatt class, which is controlled by Siemens TCS platform. The following TG equipment should be powered by the emergency diesel generator, ensuring the turbine can be safely stopped in emergency (Table 1).

The reload process of TG equipment is as follows [2, 6]:

- The equipment will stop instantly as soon as the normal bus bar fails, but the operation feedback of the equipment will remain valid, according to line A.
- The emergency diesel generator prepares to start for no more than a time after the bus failure.
- When the emergency diesel generator is ready, normal load breaker will turn off, the operation feedback of the equipment will become invalid, and the load shedding command will be active, which time is 0, 25 or 40 s, according to line B.
- After the normal load breaker off, the load switch of the emergency diesel generator will be on and the voltage of 6.6 kV bus bar will be restored, shown as line C.
- Once the load shedding command disappears, the equipment will begin to reload, shown as line D (Fig. 2).

The load shedding of TG equipment is achieved by both 1E-DCS and NC-TCS in the control process shown as follows [7]:

- Normal start-stop control logics are completed in NC-TCS.
- Load shedding command are issued from 1E-DCS, and sent to NC-TCS by redundant hard-wire.
- The priority logic of load shedding over normal start-top is realized in NC-TCS.
- After priority, the final command is transmitted from NC-TCS to MCC by hard-wire, achieving start-stop in normal operation and load shedding in emergency (Fig. 3).

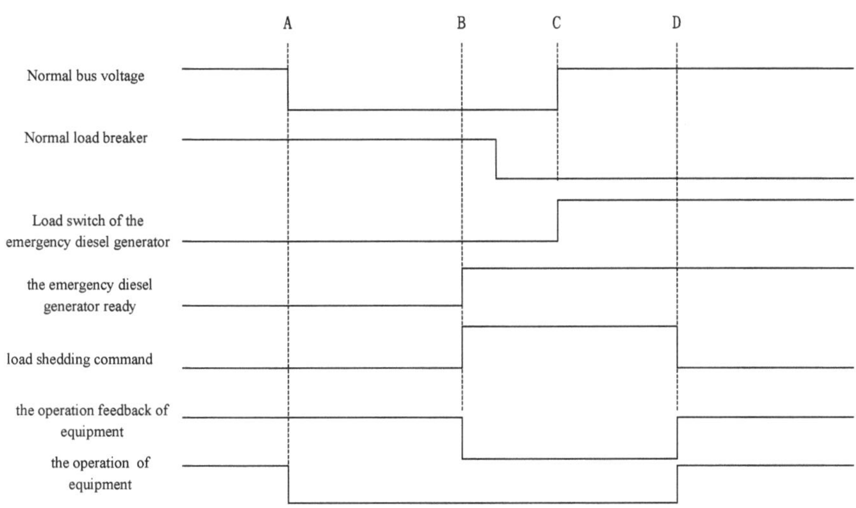

Fig. 2 The diagram of TG equipment load shedding sequence

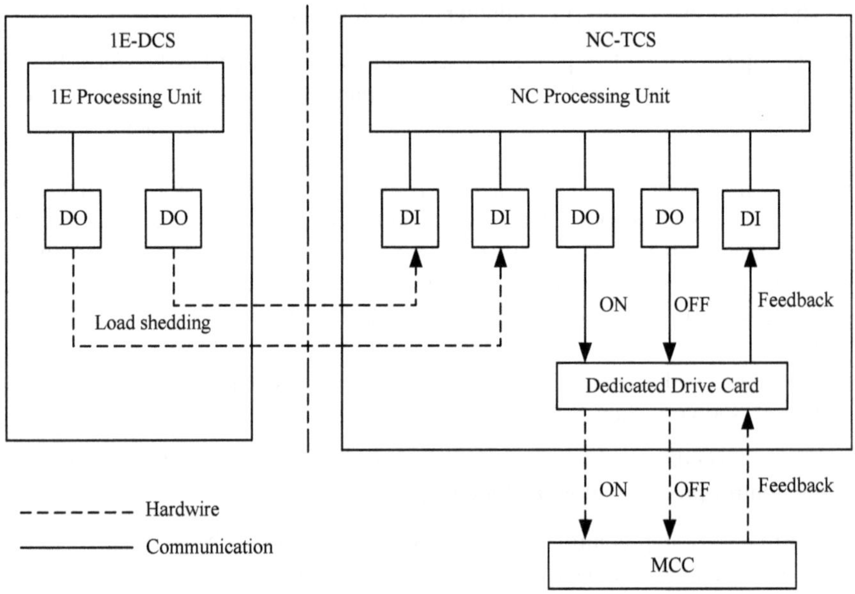

Fig. 3 The control scheme of TG equipment load shedding

The priority logic of load shedding over normal start-top is realized by the control module of sub-loop control (SLC) and DCM SWITCH in NC-TCS [5]:

- Under normal circumstances, the equipment switch module and sub-loop control module are in operation so that TG equipment can normally start up and be redundantly switched.
- When the load shedding signal is valid, the stop command comes out and the start-up command is reset, achieving the load shedding of TG equipment.
- When the load shedding signal disappears, the start-up command comes out and the fault signal of switching module is reset, achieving the reloading of TG equipment (Fig. 4).

2.2 *Analysis of TG Load Shedding Scheme*

The characteristics of TG equipment load shedding scheme are as follows:

- TG load shedding signals from 1E-DCS are two redundant hard-wired signals, which proves higher reliability.
- The priority logic of load shedding over normal start-top is realized by soft logic in NC-TCS.

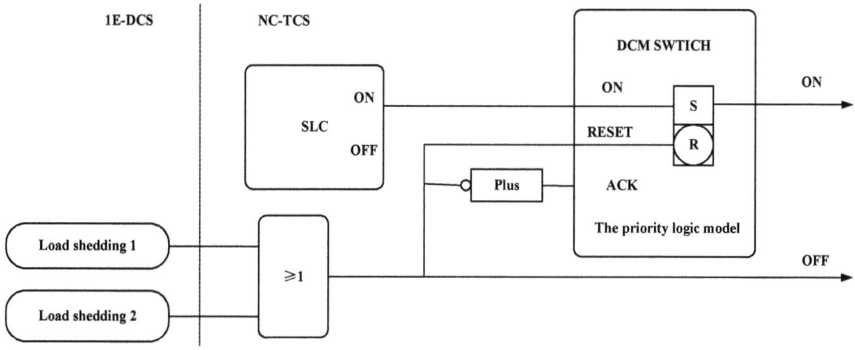

Fig. 4 The priority logic of TG equipment load shedding

Table 2 The characteristics of TG equipment load shedding control scheme

Characteristics		NPP 1
Load shedding command	Source	1E-DCS
	Redundancy	Redundant
Priority function	Platform	NC-TCS
	Implementation	Soft logic
Load shedding path	Implementation	1E-DCS → NC-TCS → MCC

- The load shedding path of TG is that: "1E-DCS sends the load shedding command → NC-TCS determines the priority → Motor Control Center starts or stops the equipment" (Table 2).

According to the analysis, the scheme of TG equipment load shedding has the following defects:

- The path of TG load shedding is not a whole 1E class path, which decreases the reliability of load shedding command. It is possible that the breakdown of the priority card or the platform in NC-TCS will result in the failure of TG load shedding command.
- The priority logic of load shedding over normal start-top is realized by not hardware in 1E-DCS, but soft logic in NC-TCS, which influences the success of TG load shedding. It is possible that the breakdown of the priority card or the platform in NC-TCS will result in the failure of TG load shedding.
- The safety function of the emergency diesel generator is affected. When it needs to start the emergency diesel generator because of loss of off-site power, if TG equipment load shedding fails, the load of the significant equipment in nuclear island may fail because the whole plant shared one diesel generator unit.

Taking into account the risk that the failure of TG equipment load shedding leads to the failure of nuclear island equipment loading, it is essential to improve TG load shedding scheme to enhance the reliability of TG load shedding.

3 Improvement of TG Load Shedding Scheme

According to the defects of the original TG load shedding scheme, the rational improvement scheme has been designed as follows:

- A new dedicated 1E class path is added for TG load shedding, which is from 1E-DCS to MCC directly by hard-wire, not through NC-TCS.
- A new priority function is realized by hard logic in the MCC control loop, not by soft logic in NC-TCS, so that the priority of new load shedding command is not influenced by NC-TCS, ensuring the success of TG load shedding.
- The logic design of TG load shedding in NC-TCS is reserved (Fig. 5).

Compared to the original scheme, the improvement scheme has the following advantages:

- TG load shedding has a dedicated 1E class path, which is not affected by the breakdown of the priority card or the platform in NC-TCS, enhancing the reliability of TG load shedding command.

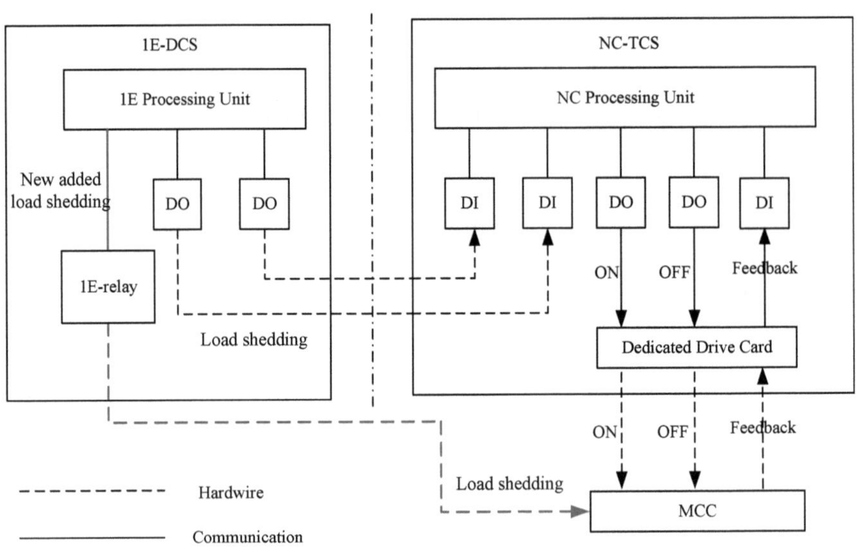

Fig. 5 The improvement scheme of TG equipment load shedding control

Table 3 Comparisons of TG equipment load shedding control schemes before and after the improvement

Characteristics		Before the improvement	After the improvement
Load shedding command	Source	1E-DCS	1E-DCS
	Redundancy	Redundant	Redundant
Priority function	Platform	NC-TCS	MCC + NC-TCS
	Implementation	Soft logic	Hard logic + Soft logic
Load shedding path	Implementation	1E-DCS → NC-TCS → MCC	1E-DCS → MCC 1E-DCS → NC-TCS → MCC

- TG load shedding has a hard priority logic in the MCC control loop, which is not affected by the breakdown of the priority card or the platform in NC-TCS, ensuring the success of TG load shedding.
- The safety function of the emergency diesel generator is not affected by the breakdown of NC-TCS, enhancing the safety of NPP.
- The logic design of TG load shedding in NC-TCS is reserved, avoiding changes of hardware interface and soft logic in 1E-DCS and NC-TCS, cutting the cost of nuclear power engineering (Table 3).

4 Conclusions

TG is the most important equipment in conventional island of NPP. On one hand, it's necessary to supply power to TG equipment when off-site power supply fails in emergency situations. On the other hand, the power supply to TG must not affect that to other more important equipment for the overall security of NPP. The scheme of TG equipment load shedding is researched in this paper, and the defect is found which might cause potential danger to the plant. The improvements were put forward to eliminate potential security risks and enhance the security. The improved scheme in this paper has been successfully applied in nuclear power projects and provides experiences for the following projects on TG equipment load shedding design.

References

1. Allen R E (1970) Standby Power Supplies for Nuclear Generating Stations. IEEE Transactions on Nuclear Science 17(1):608–615.
2. Guoqiang Zhou, Yu Qiu (2012) Analysis of the Time Influence Factors when Startup of the Emergency Diesel Generator of Nuclear Power Plant. Nuclear Power Engineering 32(z1): 31–33.

3. IEEE Standard Criteria for Diesel-Generator Units Applied as Standby Power Supplies for Nuclear Power Generating Stations (1996).
4. Moses C S, Swart P K, Dodge R E (1983) IEEE Standard Periodic Testing of Diesel-Generator Units Applied as Standby Power Supplies in Nuclear Power Generating Stations. Geochemistry Geophysics Geosystems 7(9):1–12.
5. Xiaolei Zhan, Guang Meng, Tian Deng (2013) A system and method of turbine equipment load shedding by emergency diesel generator in nuclear power plant. CN103000241A.
6. Xiusheng Lu, Zhifei Zhong, Yong Yang (2011) Study on Issues of CPR1000 NPP Emergency Diesel Generator Test. Nuclear Power Engineering 32(1):25–28.
7. Yuexin Liu, Yiming Zhong (2014) Application of None-safety Classified Priority Control Technique in CPR1000 NPP Project. Atomic Energy Science and Technology z2: 868–872.

The Research About Explosive Gas Atmosphere and Explosion Proof Technique of Waste Gas Treatment System in Nuclear Power Plant

Dan Xu, Hua Huang and Jing-Jie Bo

Abstract Hydrogen waste gas has been produced during normal operation of pressurized water reactor nuclear power plant and sent to the atmospheric environment after treatment in confined space areas. In this process, hydrogen explosive atmosphere is formed and it becomes a threat to the production safety. This article analyzed the potential release resource and explosive environment in the wastes treatment buildings, and presented a few of I&C explosion proof techniques applicable to the nuclear power plant. Commonly used explosion proof techniques of I&C design include flameproof enclosure type and intrinsically safe type. Flameproof enclosure technique focuses on the integrity while intrinsically safe technique emphasizes the compatibility of equipment installation environment and all the apparatuses in the system. The analyses show that the explosive gas atmosphere can be classified as secondary zone and application of different explosion proof techniques should take consideration of different aspects.

Keywords Nuclear power plant · Waste gas · Hydrogen · Explosive gas atmosphere · Explosion proof

1 Introduction

There are two kinds of waste gases being produced in CRP1000 nuclear power plant normal operation. One is hydrogen and the other is oxygen. Hydrogen comes from two ways. When the reactor coolant is under radiation, water can be decomposed into hydrogen and oxygen. Meanwhile, if the oxygen is over the limit, hydrogen will be injected into the chemical and volume control system for oxygen suppression. After the waste gas circulates in the reactor primary loop, it is collected into the waste gas treatment system for stocking and irradiation decaying. The

D. Xu (✉) · H. Huang · J.-J. Bo
China Nuclear Power Design Co. Ltd., Shenzhen, China
e-mail: xudan@cgnpc.com.cn

Y. Xu (ed.), *Nuclear Power Plants: Innovative Technologies for Instrumentation and Control Systems*, Lecture Notes in Electrical Engineering 400,
DOI 10.1007/978-981-10-3361-2_6

radioactivity will be checked before the gas is discharged into the atmospheric environment.

In the waste gas treatment system, hydrogen concentration can reach 70 vol.%, and this article shows how to classify the explosive area and how to choose the technique methods of explosion proof for I&C system.

2 Explosive Environment Analysis

Equipments of CPR1000 nuclear power plant waste gas treatment system include gas containers, pipes, valves, compressors and instruments. They are laid in nuclear auxiliary building which can be treated as confined area. The building takes artificial ventilation and the ventilation capacity is 0.3 m^3 per floor square meters. It meets the requirement of well-ventilated area [1].

During normal operation, potential hydrogen releasing resource can be classified as secondary source of release. Hydrogen characteristics are shown in Table 1. According to the plant operation experience, the probability of hydrogen release is quite low, and the existing time of hydrogen mixture in surrounding environment is under 1 h.

However, hydrogen concentration in waste gas treatment system is high. Once leakage happens, there will be a high hydrogen concentration area around the leak point. Production safety is threatened. So, it's reasonable to classify the explosive environment as a secondary hazardous zone.

2.1 Source of Hydrogen Release Analysis

2.1.1 Containers

Hydrogen stays in pressure release tank, volume control tank, and coolant drain tank, gas-stripper of boron recycle system, waste gas buffer tank and decay tank. Source of hydrogen release is formed when the containers' connections with outside fail.

Table 1 Hydrogen parameters

Gas	Formula	Group	T class	Ignition temp (°C)
Hydrogen	H_2	IIC	T1	500
FP (°C)	Flammability limit (V%)		Relative density	Minimum ignition energy (mJ)
	Lower	Upper		
–	4	75	0.1	0.019

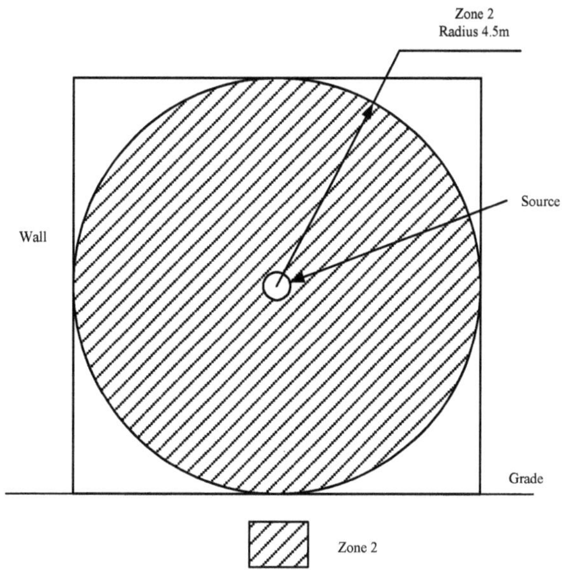

Fig. 1 Hazardous classified area

The volume of these containers is less than 95 m^3, and the pressure is under 3.5 MPa when the flow is less than 38 L/s. The area in radius of 4.5 m from the releasing source center should be classified as secondary hazardous zone [1], as shown in Fig. 1.

2.1.2 Pipes

No consideration of the possibility of hydrogen release in case of pipe breaking down.

2.1.3 Valves

The potential hydrogen releasing point is the flange connection, and the area in radius of 4.5 m from the releasing source center should be classified as secondary hazardous zone [1], as shown in Fig. 1.

2.1.4 Compressors

The waste treatment system sets two compressors, and the connection style of the gas inlet and outlet is flange. The interface of the flange's two parts is the potential hydrogen releasing point. The area in radius of 3 m from the releasing source center should be classified as secondary hazardous zone [2], as shown in Fig. 2.

Fig. 2 Hazardous classified area for compressor building

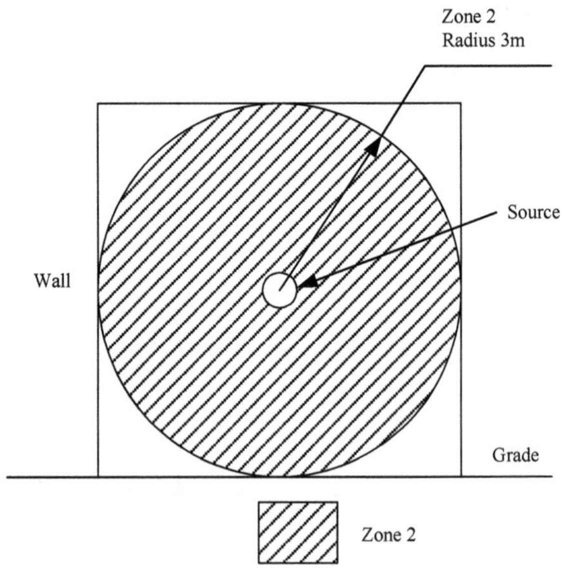

2.1.5 Instruments

The interfaces of instruments and containers or pipes whose connection style is thread or flange are the potential sources of hydrogen release. The explosive hazardous area can be classified as same as Sect. 2.1.1, as shown in Fig. 1.

2.2 Hazard Analysis of Hydrogen Area Inside Equipments

To avoid the explosion danger caused by hydrogen waste gas flows in the containers, pipes and compressors, CPR1000 nuclear plants take the measure of limiting oxygen concentration under 4 vol.%. An oxygen concentration monitor is set at the entrance of the waste gas treatment system. Once oxygen concentration reaches the upper limit, alarms will be produced and nitrogen charging system starts work.

3 Analysis of Explosion Proof Design of I&C System

The control system of waste gas treatment system of CPR1000 nuclear power plant locates in electric building, and it is separated from the site equipments in hazardous area of auxiliary building, as shown in Fig. 3. The I&C electric equipments include temperature and pressure switches, thermal couples, thermal resistance, and

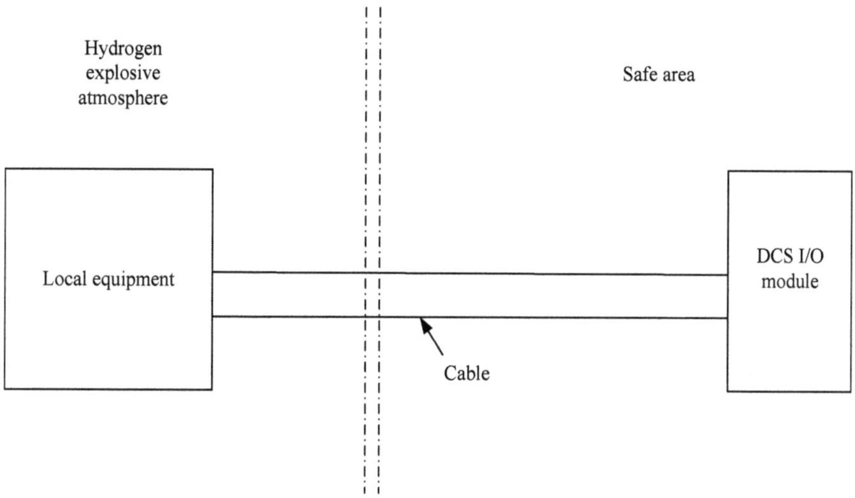

Fig. 3 Equipment connection model of nuclear plant waste gas treatment system

transmitters, analyze instruments and electric actuators of valves. The I&C control system is DCS.

The equipment protect level (EPL) adapting to secondary hazardous zone includes Ga, Gb and Gc, and all of the explosion proof methods are applicable. Flameproof enclosure method is used in nuclear power plant usually. But this method causes some inconvenience which can be solved by intrinsically safe circuit method.

3.1 Flameproof Enclosure Method

Flameproof enclosure equipment has a flameproof enclosure, in which the parts which can ignite an explosive gas atmosphere are placed and which can withstand the pressure developed during an internal explosion of an explosive mixture, and which prevents the transmission of the explosion to the explosive gas atmosphere surrounding the enclosure [3].

This method just considers the flameproof ability of equipments and related accessories in explosive gas environment while without considering the equipments in safe areas. It makes system design more simply and purchasing work more easily while equipments exchanging more difficult. For example, the length of cables supplied together with flameproof enclosure instruments should be especially required, and more junction boxes are added in the plant. Problems of cables without plenty length sometimes happen in plant installation phase. When the equipments fail to work and are taken outside of the plant to repair, cables also should be pulled out of the cable tray, and this may be a burden to plant operators.

Pay attention to the junction surface of flameproof enclosure, there should be a 40 mm distance between the surface and surrounding barrier. Enclosures are not allowed to open or structure changed casually. Operators should judge the maintenance opportunity according to the plant hydrogen detection system operation condition.

3.2 *Intrinsically Safe Circuit Method*

Explosion proof method of intrinsically safe circuit is based on the restriction of electrical energy within equipment and interconnecting wiring exposed to the explosive atmosphere to a level below that which can cause ignition by either sparking or heating effects [4]. Normally intrinsically safe equipments are integrated together to form an intrinsically safe circuit in engineering.

Intrinsically safe circuit limits the energy of the control system from the design point so that the energy is less than the minimum igniting energy of explosive gas and the maximum surface temperature is below the igniting temperature of the explosive gas. As shown in Table 1, the minimum igniting energy of hydrogen is 0.019 mJ, igniting temperature is 500 °C, and the allowed maximum surface temperature is 450 °C. Since only different parts of the whole system match well can an intrinsically safe circuit system play its function, the whole system is purchased rather than single equipment.

The following method shows a way for judging the intrinsically safe circuit of simple system.

3.2.1 Equipments in Hazardous Area

1. Thermal resistance and switches: belong to simple equipment.
2. Thermal couple: the waste gas temperature is around 100 °C and usually K type, and electromotive force is 5 mV. The maximum current of DCS I/O module is 50 mA. So the thermal couple can be treated as simple system as well.
3. Transmitters, solenoid valves and others: should be specially identified.

3.2.2 Safety Barrier

Intrinsically safe DCS I/O modules haven't been used in the CPR1000 nuclear power plants, and it's more convenient to use safety barrier in intrinsically safe circuit design than develop a new I/O module.

3.2.3 Cables

The length of cables can be known once the places of start equipment and end equipment are fixed. So the distributed inductance and capacitance value of cables can be calculated by referring cables' characteristics. Equivalent model of cable distributed parameters are shown in Fig. 4 [5].

The cable length can reach 120 m from the equipment in hazardous area to DCS cabinet in electric building. The transient voltage and current which is produced by closing and opening circuit can be calculated by referring the model in Fig. 5 [5], and the energy shouldn't exceed 19 μJ.

3.2.4 System Assessment

When the signal collection and equipment driver both take single power supply, simple analyze method can be used to assess the intrinsically safe circuit. The model is shown in Fig. 6.

The following equations should be satisfied [6]:

$$U_o \leq U_i$$

$$I_o \leq I_i$$

$$P_o \leq P_i$$

$$L_c + L_i \leq L_o$$

$$C_c + C_i \leq C_o$$

Requirements for load resistance:

$$R_c + R_i + R_o \leq R_d$$

(R_d is the allowed external resistance of DCS I/O module) [7]

There are special requirements for cable identification in nuclear power plant, and consideration of both cable special identification and intrinsically safe requirement should be taken. Intrinsically safe circuit uses single-point grounding in safe areas, or equal potential grounding in hazardous areas.

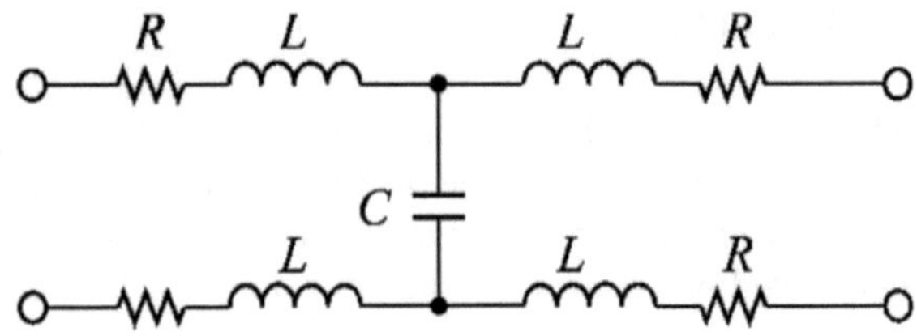

Fig. 4 Equivalent model of cable distributed parameters

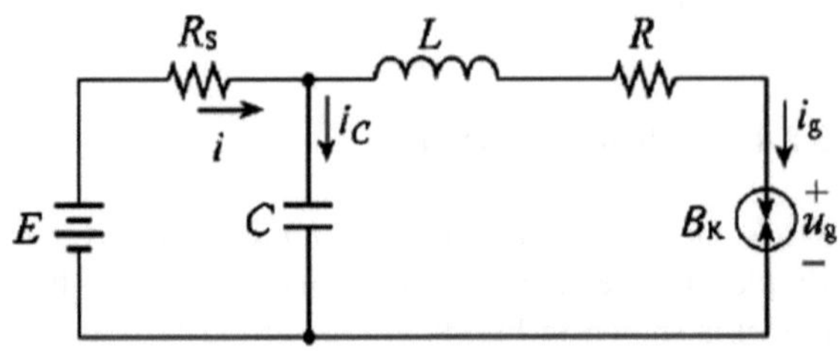

Fig. 5 Cable circuit model under oscillation condition

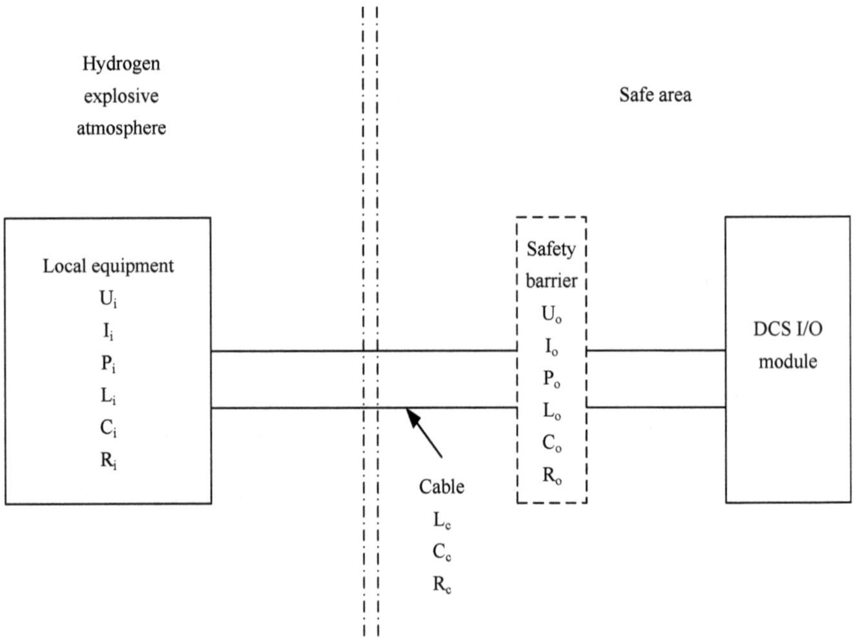

Fig. 6 Equipment connection model of intrinsically safe electrical system

4 Conclusions

1. The hazardous areas belonging to waste gas treatment system of CPR1000 nuclear power plant can be classified as secondary zone;
2. Flameproof enclosure technique is easier to carry out, but corresponding maintenance is inconvenient. Intrinsically safe technique can avoid unnecessary equipments' explosion proof requirements by analyze and it's much safer than other explosion proof methods, but it's quite complicated for the whole system design;
3. Consideration of system scale and purchase willing should be taken to choose the explosion proof methods.

References

1. GB 50058-2014 Code for design of electrical installations in explosive atmosphere.
2. SY/T 6671-2006 Recommended practice for classification of locations for electrical installations at petroleum facilities classified as class I, zone 0, zone 1, and zone 2.
3. IEC 60079-1-2007 Explosive atmospheres-part 1: Equipment protection by flameproof enclosures "d".
4. IEC 60079-11-2006 Explosive atmospheres-part 11: Equipment protection by intrinsic safety "i".
5. SUN, J. P., WANG, Z. L., LIU, X. Y., ZHANG, L. Y., & CUI, L. Z. (2010). The distributed parameters and discharge characteristic of communication and signal cable. Journal of China Coal Society, 7, 040.
6. IEC 60079-25-2010 Explosive atmospheres-part 25: Intrinsically safe electrical systems.
7. Li, S.L. (2000). The application of safety barrier in intrinsically safe electrical systems. Automation in Petro-Chemical Industry, (1), 6–7.

Analyzing and Processing the Malfunction of Control Valve Dithering in PWR Nuclear Power Plant

Chu-hao Xi, Long-qiang Zhang and Yong Tian

Abstract As digital technology is more and more used in nuclear power plants, the electromagnetic compatibility issues between new digital device and the plant environment are increasingly prominent. This paper is aiming at the dither of control valve in pressurized water reactor project, which is causing by the electromagnetic compatibility issues. And this paper mainly describes the experience of working process and the solving program of failure eliminating. First it introduces the installation condition and operating principle of the valve, then it analyzes the possible causes by using fault tree and find the root cause by site investigation. The root cause of the malfunction is that the feedback signal of valve position is disturbed by an interferential signal about 32 kHz. At last, this paper provides two kinds of program to solve the problem, one is grounding directly and another is grounding with capacitance.

Keywords PWR · Dithering of valve · Fault tree · Ground · Electromagnetic compatibility

1 Introduction

As digital technology is widely used in nuclear power plants, the electromagnetic compatibility issues between new digital device and the plant environment are increasingly prominent. Because the working frequency of digital technology is higher than the analog technology, which contributes to extend the band width of interference and increases the risk of being disturbed. Therefore, it raises some new questions. The fault of certain type of control valve with digital positioner was found in the debugging process of some PWR power plant, the fault reason is largely related with electrical interference. This paper describes the working process and solutions in dealing with the above-described engineering problems.

C. Xi (✉) · L. Zhang · Y. Tian
China Nuclear Power Plant Design Co., Ltd., Shenzhen, China
e-mail: xichuhao@cgnpc.com.cn

Y. Xu (ed.), *Nuclear Power Plants: Innovative Technologies for Instrumentation and Control Systems*, Lecture Notes in Electrical Engineering 400,
DOI 10.1007/978-981-10-3361-2_7

2 Background

In the Debugging phase of a PWR power plant project, when a certain type of valve is receiving the input signal from a portable signal generator, there is no significant abnormity about valve and positioner. But when that valve is receiving the input signal from DCS, it found that this type of valve position is obvious dithering, at the same time the output pressure of valve driver exists obvious fluctuations too. And when the debugging staff re-runs the self-tuning function, it was found unable to successfully complete the self-tuning function (after about 15 min the tuning has failed). After observation and analysis, above phenomenon shows that this type of valve may be under interference and need to be eliminated.

3 Failure Analysis

3.1 Valve Principle

The control channel of the valve is installed as shown in Fig. 1, DCS output 4–20 mA control command that send to the positioner though local CR box. The positioner and position sensor form a closed control loop. The positioner receiving setpoint from DCS. At the same time, it acquires valve position feedback signal from sensor, then after internal PID operation output signal to drive the actuator control valve actuation.

Positioner electronic module consist of a multiplexer, an A/D converter, a D/A converter, a temperature sensors, a Hall-effect position sensor (or a slide rheostat), a pressure sensor, a microprocessor and a power distribution management circuit. The diagram of functional block linking shown in Fig. 2 [1, 2].

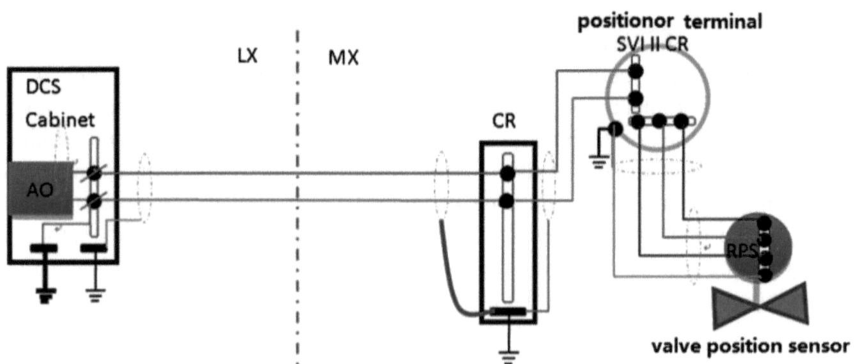

Fig. 1 Control channel installation diagram

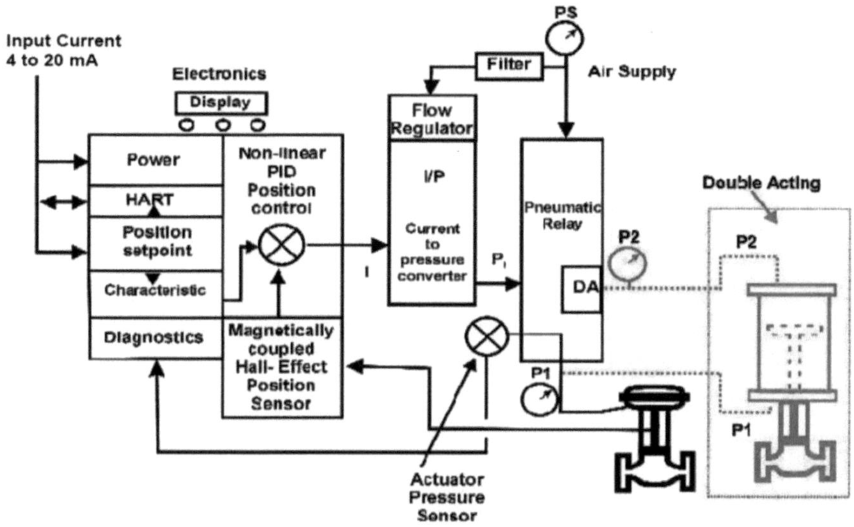

Fig. 2 Block diagram of internal positioner

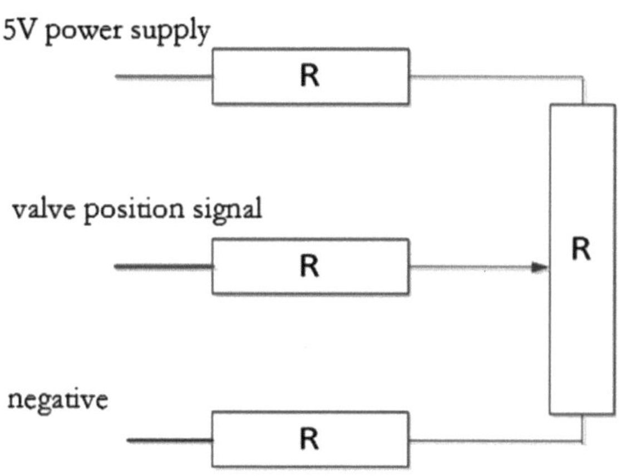

Fig. 3 Feedback signal of valve position

The feedback signal of valve position is provided by a circuit of sliding resistance, as it is show in Fig. 3. The slide terminal will move as the valve position, thereby changing the feedback signal of valve position.

3.1.1 Fault Tree Analysis

By using Fault tree [3, 4], it gradually split the reasons though the following two possible path. The establishment of the fault tree shown in Fig. 4. The possible theoretical reasons were as following section.

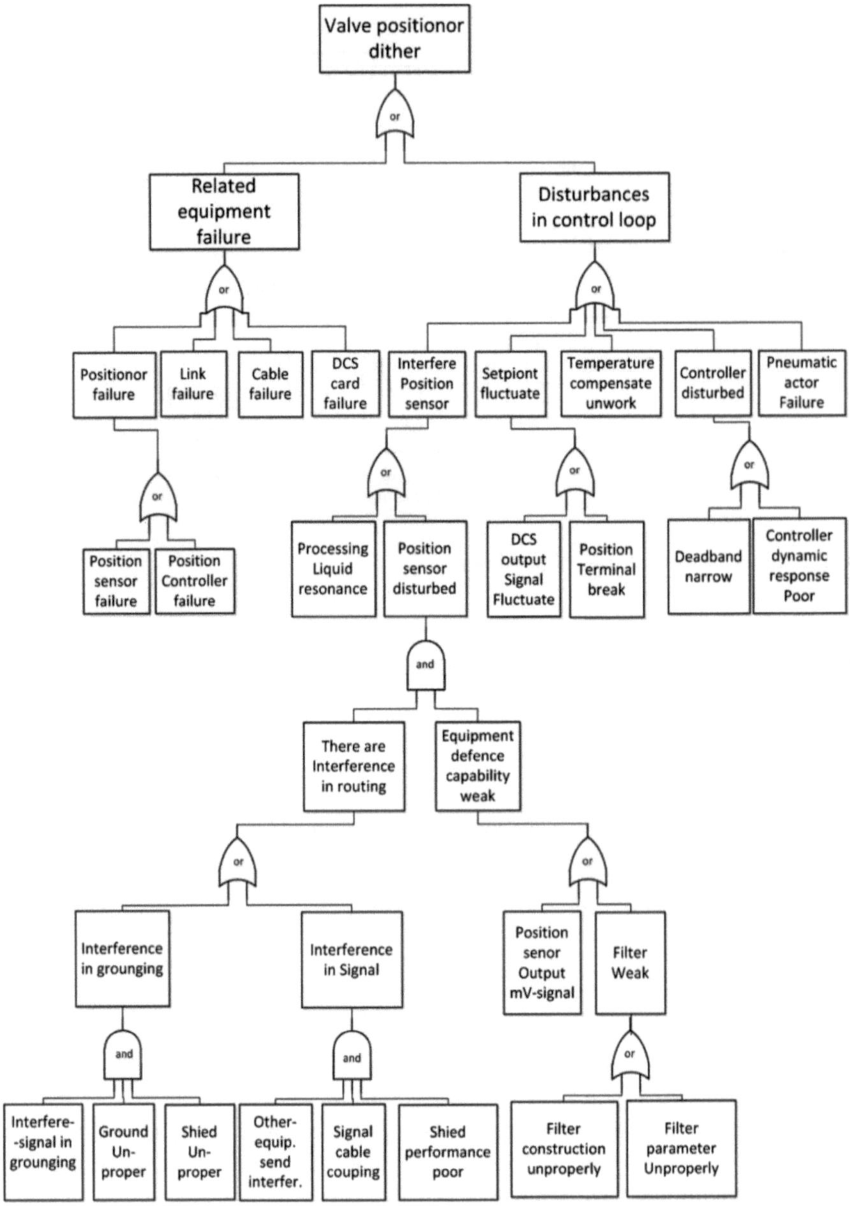

Fig. 4 The fault tree analysis diagram of positioner dither cause

Related Equipment Failure

Valve dither is due to equipment failure caused by:

1. The failure of positioner including two aspects:
   - the failure of controller may cause the output control command abnormal, causing the valve dither.
   - the failure of valve position sensor could cause the controller getting unstable or in accurate feedback signal of the valve position, causing the valve position dither.
2. The failure of linking

Equipment installation at the connection of substandard, may form a non-constant contact resistance. In the field the devices inevitably have varying degrees of shock or vibration, so it may cause the non-constant contact resistance changing, therefore the valve dither.

3. The failure of cable

Cable insulation resistance or shield is under substandard. It may bring in interference signal, so the valve dither.

4. The failure of DCS card

The failure of DCS analog output card cause the substandard of the output signal. So the supply voltage of positioner is instability, causing the valve dither.

Interference in Control Loop

Valve dither due to the interference in control loop:

1. Interference to set point, including two aspect:
   - DCS output control command fluctuating.
   - The performance of positioner's receiving terminal is under substandard, resulting setpoint instable.
2. Interference to valve position sensor, including two aspect:
   - valve position sensor is interfered by turbulence of the process fluid.
   - valve position sensor is affected by electromagnetic interference.
3. Interference to temperature compensation circuit.

A temperature sensor is positioned within the electronic module for measuring the temperature of environment to compensate valve position signal and pressure signal.

4. Existing Problems in positioner controller design:
   - The design of control parameters or the dynamic response of control loop is under substandard.
   - The deadband of controller is setting unreasonable.
5. Interference to valve pneumatic actuator, including the pressure of gas source instability, the fluctuation of pressure amplifier's output and other factors.

3.1.2 Root Cause

Root cause is obtained by combining theoretical reason of fault tree analysis, the pre-trial test and follow-up supplementary tests [5]. The root cause is found that the terminal of 5 V power supply has a interference signal of high-frequency, and the conduction path of interference signal is shown in Fig. 5. So the feedback signal of valve position is disturbed by the high-frequency interference.

The feedback signal of valve position was measured, the waveforms is measured as Fig. 6. Obviously, output sampling signal of valve position exists high-frequency electromagnetic interference. the valve positioner signal obtained a series of fluctuations value, so the controller is receiving a series of instable values, so it send the false command to the actuator. Therefore it finally cause the valve dithering.

3.2 Failure Elimination

According to the root cause described in the preceding part of the paper, two kinds of schemes were designed to eliminate the disturbance.

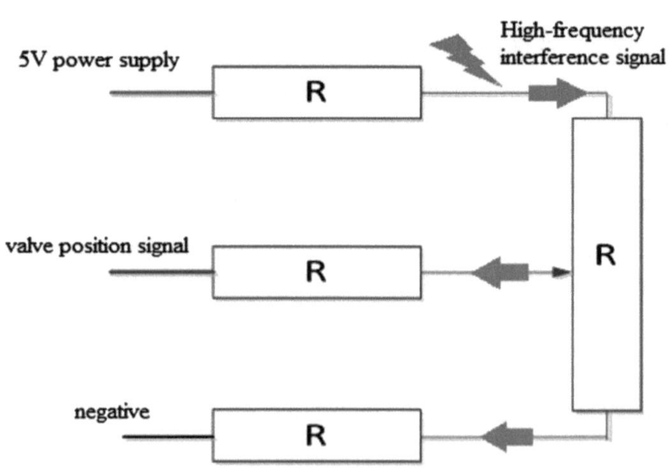

Fig. 5 The conduction path of interference signal

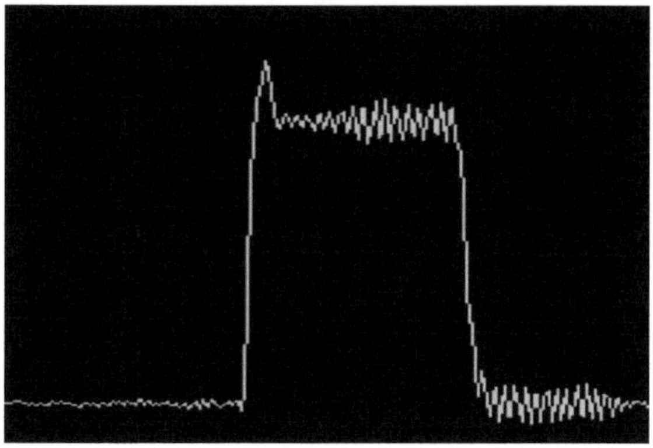

Fig. 6 The valve position feedback sampling signal waveform

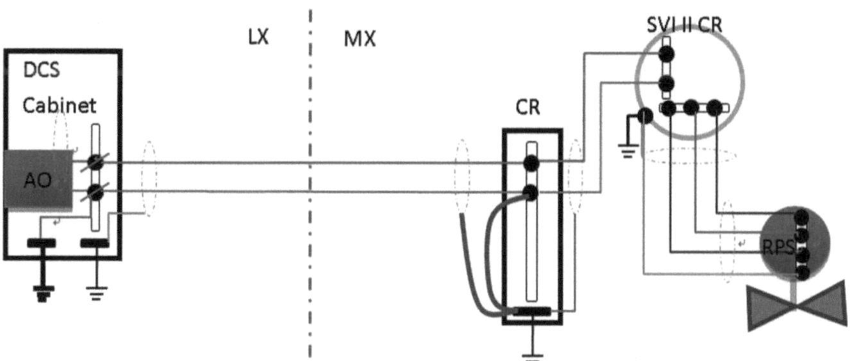

Fig. 7 Directly grounded wiring diagram

3.2.1 Direct Grounding Scheme

To eliminate the dither, the negative of signal should be linked to the ground at the CR box, the wiring diagram is shown in Fig. 7 [6, 7].

By using this scheme, we can find that the valve dither is reduced, but not completely eliminated. Because this scheme does not directly act to the root cause.

3.2.2 Capacitance Grounding Scheme

Aiming at the feature of the interference signal, a capacitor which is 1μF is added between the negative terminal and the shield grounding of the valve's position sensor to filter the interference signal, and it is show in Fig. 8. Because the impedance of capacitance is much lower than the resistance for the high-frequency interference signal. So the capacitor provide interference signal with a conduction

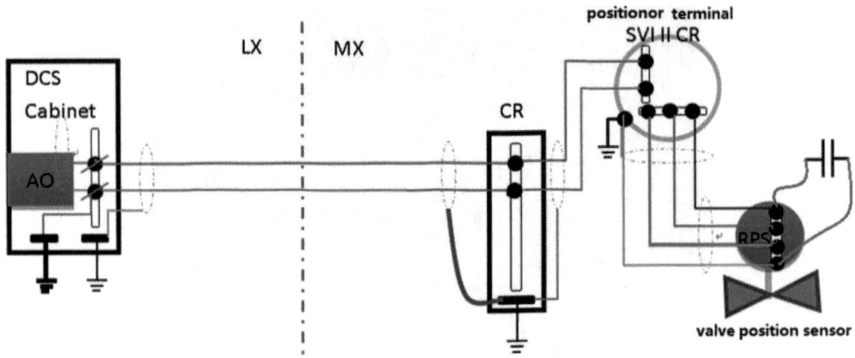

Fig. 8 Capacitance grounding diagram

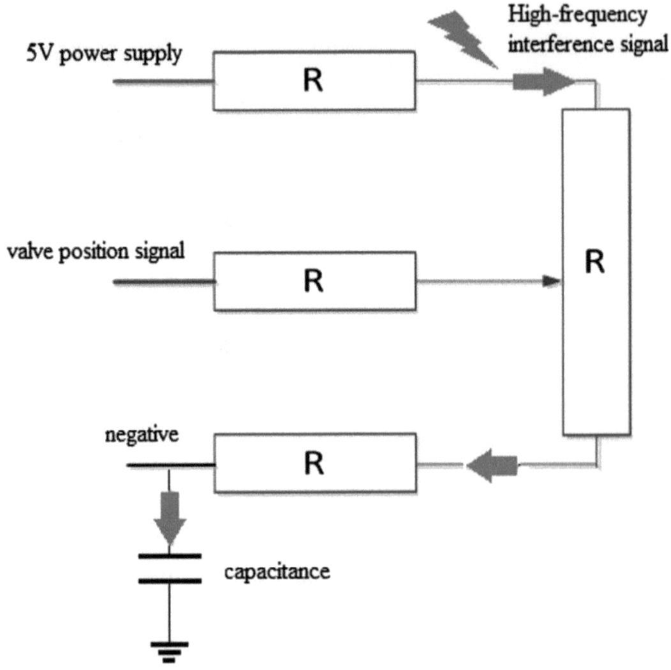

Fig. 9 The conduction path of interference signal with capacitance

path of low impedance. And the conduction path of the high-frequency interference signal will change as Fig. 9 shows. Then we could find that the valves dither is disappeared, the function of self-tuning recover and the valves work normally again.

The wave form of the valves' position sampling feedback sampling is as shown in Fig. 10 [8]. Apparently showed in the figure, the signal is smoothing and the

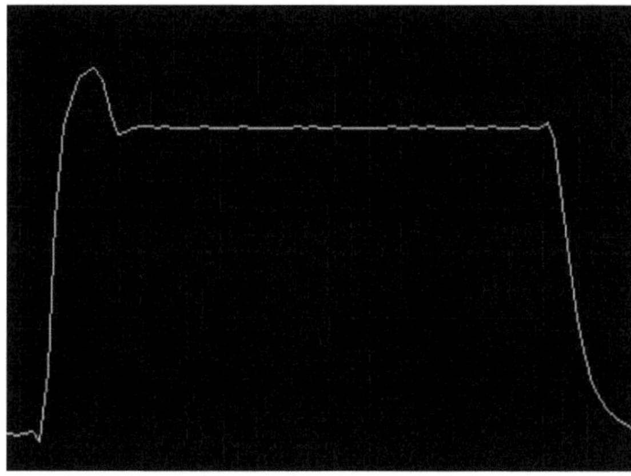

Fig. 10 The waveform of valve position sensor after adding capacitance

valves' position is back to normal. Compared to the scheme of the signal's positive is grounded directly, adding a capacity is better in restraining the jitter. We can make a conclusion that adding a capacity is more effective in filtering the interference.

Apparently showed in the figure, the signal is smoothing and the valves' position is back to normal. Compared to the scheme of the signal's positive is grounded directly, adding a capacity is better in restraining the jitter. We can make a conclusion that adding a capacity is more effective in filtering the interference.

3.3 Conclusions

This paper is aiming at the dither of control valve in pressurized water reactor project. Firstly, analyzing the root cause by using the fault tree. Then finding the root cause according to the investigation at the field and discussing two solutions' difference and the effect in filtering the interference. Finally making a conclusion that adding a capacity between the negative terminal and the shield grounding of the valve's position sensor is more efficient.

References

1. Zhou Hongyan (2001) Research on intelligent electric valve positioner. Sichuan University.
2. Zhang Lei (2001) Research on intelligent valve positioner. Luoyang Institute of Technology.
3. Chen Dedao (2015) An huping Numerical Control Machine Fault Diagnosis Based on Fuzzy Fault Tree. J Machine Tool & Hydraulics 177–178.
4. Pan Bo, Jiang Tongmin (2012) The structure of the corrosion damage of aircraft Fuzzy comprehensive evaluation based on fault tree. J Beijing University of Aeronautics and Astronautic 38–42.

5. Wu Weizhi, Cheng Shengchang, Yuan Wenzhong (2012) Power Plant Unit 4 DCS system analog signal processing analysis of wide fluctuations. J Thermal Power Generation.
6. Wang Qinghai (2012) PLC control system grounding jamming technology. J Automation, 173–174.
7. Machang (2009) Talking about the PLC control system grounding jamming technology. J Dragons Technology and Management.
8. Wang Qingliang (2012) Single-phase ground fault analysis and line selection. M China Electric Power Press.

First Loop Loose Parts Field Diagnosis

Hui-Tao Yang and Jin-Bo Zhao

Abstract This paper from the loose parts of the harm, combining with the actual situation of the field, for first loop loose parts of the alarm signal after the scene recognition and diagnosis are introduced, respectively used in the diagnostic field listen to the sound of needle method, field data playback listening method and data analysis of linear positioning method, circle intersection method, hyperbolic positioning method, Newton iterative positioning method, contrast and analysis the advantages and disadvantages. And combined with field loose off parts of echo, the reactor internal structure characteristics, equipment operation state, through the positioning analysis method to determine loose parts quality and position, of first loop loose parts field diagnosis put forward own views, for follow-up of new project a loop loose parts detection system design reference.

Keyword Loose parts · First loop · Diagnosis

1 Introduction

First loop system as an important part of the nuclear power plant, there are a large number of screws, pins and other parts of the interior, although at the beginning of the design of the anti-loosing measures, but in first loop under the continuous impact of high-speed flow and high pressure steam, these parts will be loose or even fall off. According to GB/T11807-2008 [2], including the shedding parts, loose parts and foreign object. Shedding parts refers to the reactor coolant pressure boundary components without the connection and can be carried by the coolant. Loose parts are loose, but fixed with the original parts which can still keep the

H.-T. Yang (✉) · J.-B. Zhao
Shenzhen China Nuclear Power Design Co., Ltd.,
Shenzhen 518045, Guangdong, China
e-mail: huitaoyang@163.com

J.-B. Zhao
e-mail: zhaojinbo@cgnpc.com.cn

Y. Xu (ed.), *Nuclear Power Plants: Innovative Technologies for Instrumentation and Control Systems*, Lecture Notes in Electrical Engineering 400,
DOI 10.1007/978-981-10-3361-2_8

connection parts. Foreign object refers to the reactor shutdown (e.g. refueling and maintenance) during the period of the reactor coolant pressure boundary, does not belong to the internal structure of objects. Loose parts will affect the reactor running stability and reliability; even threaten the safety of nuclear power station. In history, there are a lot of nuclear accident is due to the presence of loose parts, for example, in 1979 the American prairie island 1# units because of coil spring fall cause steam generator tube rupture; in 1982 Wisconsin cusp beach, the steam generator heat transfer guide tube leakage caused by type C clamp and other loose parts; in recent years, in the construction of a nuclear power plant project occurred during fuel assembly pin loosening caused by hot test was interrupted.

So as soon as possible in order to detect the loose parts in nuclear power plant, avoid or reduce the occurrence of loop system of safety related damage or failure, in improving the monitoring capability of loose parts monitoring system and at the same time, how in the complicated field conditions, existing technology and analysis means of reasonable use, early detection and determine the loose parts of size, location and range of influence, to provide reference data for reliable operation in the construction unit, it is particularly important.

2 Listen to the Sound Identification Method

During the normal operation of the reactor, first loop system has three main voices: structure sound, background noise, and constant background noise [2]. Structure sound refers to the propagation of sound. Propagation in solid voice in GB/T 11807-2008 [2], the structure mainly refers to the propagation of sound frequency from 30 to 20 Hz voice. The background noise refers to the structure of transmission during reactor operation, the noise in the loose parts will be. The background noise comprises a constant background noise and run single sound event. The constant background noise is caused by the random part (such as the coolant flow noise) and determining part (such as noise depends on the reactor coolant pump speed). Through the compression device between the inner component and the pressure vessel, the vibration can be transmitted to the acceleration sensor or displacement sensor [1].

Listen to the sound identification method mainly uses the structure of sound loose parts impact pressure boundary generated. As long as the loose parts, shedding parts or foreign objects impact the inner surface of the reactor coolant pressure boundary or its internal structure, energy will be transferred to the pressure boundary wall, resulting in a single sound event. The single event from dropping off the impact of the loose parts and run in single event sound frequency and loudness will have some differences; so as to identify the loose parts has laid the basis of distinguishing.

2.1 *Listening to the Sound of Needle*

Listening to the needle identification method is mainly applicable to the commissioning period. A single voice after the incident, the alarm will send to the loose parts monitoring system of main control room or cabinet, supervision personnel according to the operating conditions of equipment and components on the alarm to do the basic analysis eliminate interference due to equipment start stop or other construction site caused by. Then notify the instrument control personnel to analyze and judge. During the cold test and hot test (no fuel), because of loose part monitoring system is still in the debugging stage, site condition is complex, on-site technical personnel often used to judge loose parts of the location with needle, the main monitoring position: heat steam generator bottom hot side and cold side, the bottom of pressure vessel and between heat pipe and connecting pipe. Listening to the needle has the advantages of avoiding the false alarm, leading to the technical problems of loose parts monitoring system itself, such as sensor terminal error etc. Can more directly on the removal of parts of the landing position and the regional judge.

2.2 *Loose Part Monitoring Data Playback Sound*

In the two generation and the three generation reactor, loose parts monitoring system has data storage and playback function. During normal operation, the operator can periodically or according to the needs of listening and playback recent noise signal, at the same time according to the normal operation of change of background noise adjust peak factor, to make sure the loose part monitoring system in good condition monitoring. Familiar with the various characteristics of noise equipment start and stop the noise, improve the recognition ability.

Listen and choose the data playback to the corresponding region according to the running state of the equipment, but also according to the direction of movement of refrigerants in order to monitor playback.

3 Data Analysis Method

The data analysis method is mainly based on data collected from loose sound components; the data is calculated to determine the location, size and direction of movement of loose parts [3]. The principle of the CPR reactor type in the loose parts monitoring system and EPR reactor type in the loose parts monitoring system is the same, but in the EPR reactor, the loose part monitoring system configuration calculation mature analysis software, improves the efficiency and accuracy of field diagnosis analysis.

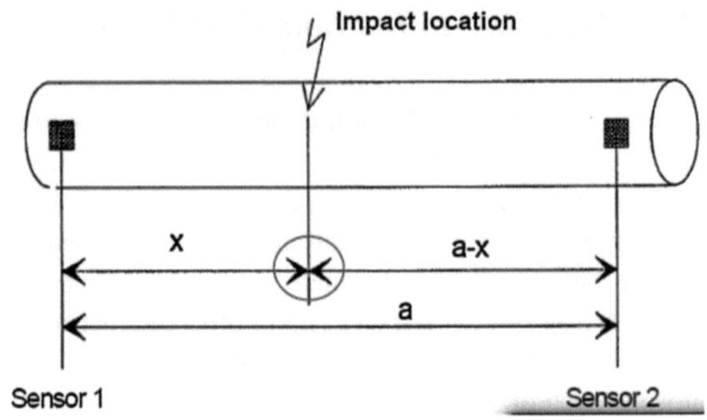

Fig. 1 linear localizing method

3.1 Linear Positioning Method

See Fig. 1.

The following calculation model of linear positioning method:

$$\Delta t = \frac{x}{c} - \frac{a-x}{c} \Rightarrow x = \frac{a - c * \Delta t}{2}, \tag{1}$$

Among them:

- Δt The measured time difference between the impact signals for the sensor 1 and sensor 2;
- a The distance between the sensor 1 and sensor 2;
- c The speed of vibration wave propagation in a pipe, different material pipes or equipment vibration wave propagation speed is different.

Linear positioning method is a kind of time difference method, only the monitoring data of two points, for the smaller diameter of straight pipe equipment. In the EPR reactor, due to hot and cold pipe and the steam generator are installed only two sensors, so the location method commonly used in the calculation of the heat pipe section and positioning in the steam generator of the parts and loose the impact point.

3.2 Hyperbolic Positioning Method

See Fig. 2.

The following calculation model of hyperbolic intersection method:

$$\begin{aligned} r_2 - r_1 &= c\Delta t_{1,2}, \\ r_3 - r_1 &= c\Delta t_{1,3} \end{aligned} \tag{2}$$

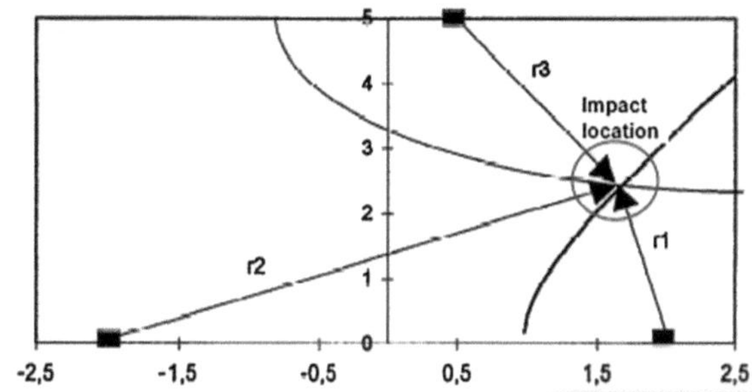

Fig. 2 Hyperbolic intersection method

r_1, r_2, r_3 Respectively distance difference between the points of impact and the three known points;
c Transmission speed of vibration wave;
$\Delta t_{1,2}$ The time difference of vibration wave between sensor 1 and sensor 2;
$\Delta t_{1,3}$ The time difference of vibration wave between sensor 1 and sensor 2.

Hyperbolic positioning method needs three monitoring data of two-dimensional calculation method, in actual calculation, and the two-dimensional coordinates of a known transducer, just assume the impact point coordinates. Based on the model of the formula established two-dimensional coordinate equation can be calculated impact point coordinates. The coordinate value is two, the mistake one can be eliminated according to the internal coordinate range.

3.3 Circle Intersection Positioning Method

See Fig. 3.

The following calculation model of circle intersection method:

$$r_1 = \frac{c_a c_s \Delta t_1}{c_s - c_a}; \quad r_2 = \frac{c_a c_s \Delta t_2}{c_s - c_a}, \tag{3}$$

c_a Symmetric wave speed;
c_s Asymmetric wave velocity;
Δt_1, Δt_2 The time difference of two different sensors monitoring points.

Circle intersection positioning method is realized by calculating the absolute distance between position of sensor and impact point, using the characteristics of the bending wave propagation, velocity in the propagation process of different frequency of flexural wave is different, with the increase of propagation distance, various frequency flexural waves separated gradually. In the waveform, wave crest and wave front be open and wave energy attenuation. In theory, as long as more than two sensors can realize the positioning, estimate the collision point of absolute

Fig. 3 Cycle intersection method

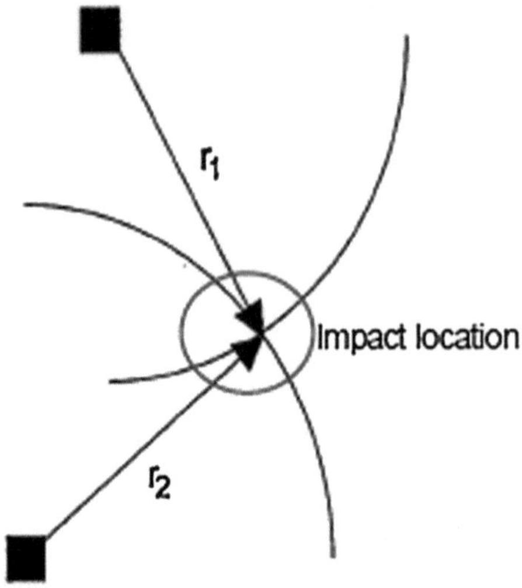

distance as the radius, sensor position as the center circle, then the circle intersection point is the loose parts impact position. The installation position of sensor is fixed, so the key of this method is to determine the sensor absolute distance to the point of impact, the commonly used methods for estimating the absolute distance measurement, the peak vibration modal method, time-frequency analysis method, reference collision method etc. But these methods have a common problem, namely cannot determine the accurate position of loose parts impact, can only determine the approximate position.

3.4 Newton Positioning Method

See Fig. 4.

The following calculation model of Newton method:

$$\begin{aligned} & d_i = c\Delta t_i; \\ & p_i = l_i - l_0; \\ & p_i = \sqrt{(x_i - x_s)^2 + (y_i - y_s)^2} - \sqrt{(x_0 - x_s)^2 + (y_0 - y_s)^2}; \\ & Q = \sum_{i=1}^{n} (d_i - p_i)^2 \end{aligned} \tag{4}$$

Among them:

c Transmission speed of vibration wave;

d_i Distance difference between the points of impact and the two known points;

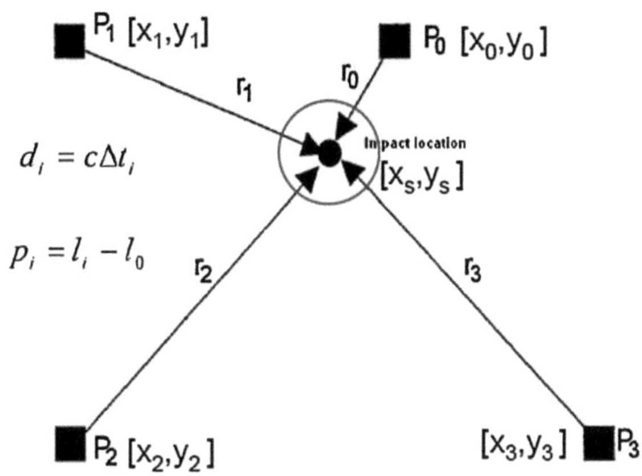

Fig. 4 Newton (gradient) method

Δt_i The time difference of vibration wave between $[x_i, y_i]$ and $[x_0, y_0]$;
p_i The distance difference of vibration wave between $[x_i, y_i]$ and $[x_0, y_0]$;
n The number of time difference.

Newton method is a general algorithm for linear calculation, two-dimensional calculation and dimension calculation, in the positioning calculation plays an important role, but the Newton iterative calculation process is complex, and with the increase of the number and dimension involved in the calculation, the amount of calculation increases exponentially. Frequently used in computer programming. In the EPR reactor type loose part monitoring system, Newton iterative method is applied for the calculation of steam generator, heat pipe segment and RPV and other parts of multiple points.

4 To Estimate the Quality of Loose Part

In the field of diagnosis process, the quality of the loose parts estimation mainly through the following ways:

1. By tapping on the primary pressure boundary and the size of the voice of artificial estimation.
2. Through the analysis of the reactor components may be loose to estimate, but due to the possibility of external components, the estimation method of errors is very high.
3. By calculating software to calculation and analysis, EPR type of reactor loose parts monitoring system equipped with quality evaluation software, through the loose part of the pressure boundary impact strength estimated the size of loose parts, but due to vibration and equipment operation noise of reactor device, the author analysis the estimated the accuracy is not high.

So to improve the efficiency of quality estimation and the accuracy of loose parts quality, need to establish internal loop pressure boundary especially in internal component of every parts of the shape, quality, location of the database, improve the ability of analyzing and filtering background noise, avoid impact signal missing low mass of loose parts, the accuracy of analysis software and improve the level of automation.

5 Conclusions

Starting from the concept of safety accidents and loose components caused by the loose parts, expounds the importance of loose parts monitoring system, combining with the current technical level of loose part monitoring system, based on the actual conditions are summarized and analyzed six kinds of field diagnostic method, respectively used in the diagnostic field listen to the sound of needle method, field data playback listening method and data analysis of linear positioning method, circle intersection method, hyperbolic positioning method, Newton iterative positioning method, contrast and analysis the advantages and disadvantages. And combined with field loose off parts of echo, the reactor internal structure characteristics, equipment operation state, through the positioning analysis method to determine loose parts quality and position, for follow-up of new project a loop loose parts detection system design reference, has a certain guiding significance on the loose parts of site diagnosis.

References

1. EJ_T 1188 (2005) Vibration monitoring of internal components of pressurized water reactor in nuclear power plant.
2. GB/T 11807 (2008) The characteristics, design and operational procedures of acoustic detection system and exploration of loose parts.
3. Pei-Gen Zhang (2012) Analysis of the key technology of loose parts monitoring system (C). The 2012 nuclear power plant digital control seminar 10:176–178.

Research on Application of Remote IO Technology in Nuclear Power Plant

Hui Wang, Tian-Sheng Ji, Ai-Guo Lv and Hao Zhong

Abstract The technology of remote I/O system has the characteristics of good physical dispersion, high reliability, high accuracy of measurement and control, openness, strong function, high degree of intelligence and strong communication ability. Remote I/O technology, as a mature bus technology, has been widely used in thermal power, chemical industry, metallurgy, petroleum and other fields. All systems and sub items in nuclear power plant are still using the traditional distributed control system (DCS), some signals by hardwire have a long distance transmission, which against economic and signal stability. In this paper, from the plant and system environment restriction condition, functional requirements (Operational protection requirements), system performance requirements, installation requirements, equipment identification, cost and other system interface and other factors, to analyze the feasibility of remote I/O technology used in nuclear power plant. Determine the scope of the implementation of remote IO technology for nuclear power plant. Finally, pumping station as an example, the implementation plan of using the remote I/O technology under the various technical route of nuclear power plant was studied in detail.

Keywords Remote I/O technology · Distributed control system (DCS) · Local control system · Pumping station

1 Introduction

The remote I/O technology transfers the data processing and control function of the traditional DCS to the field equipment nearby, and the multiple monitoring points are connected with the PLC or DCS through the bus. The remote I/O technology have been widely used in thermal power, chemical industry, metallurgy, petroleum

H. Wang (✉) · T.-S. Ji · A.-G. Lv · H. Zhong
State Key Laboratory of Nuclear Power Safety Monitoring Technology and Equipment, China Nuclear Power Engineering Co., Ltd., Shenzhen, China
e-mail: hui.wang@cgnpc.com.cn

Y. Xu (ed.), *Nuclear Power Plants: Innovative Technologies for Instrumentation and Control Systems*, Lecture Notes in Electrical Engineering 400,
DOI 10.1007/978-981-10-3361-2_9

and other fields, the technology has been very mature. But all nuclear power plant system and sub items still use the traditional DCS. Some signals were long-distance transmission, detrimental to the economy and the stability of the signal, it is necessary to research and analyze the feasibility of using remote IO in nuclear power plant, and timely improved.

The technology of remote I/O system has the characteristics of good physical dispersion, high reliability, high accuracy of measurement and control, openness, strong function, high degree of intelligence and strong communication ability. At the same time, the economic benefits are obvious.

2 The Application of Remote IO in Conventional Thermal Power Plant

Remote I/O system has been widely used in thermal power plant, the application of the main two aspects of thermal power plant.

2.1 *To Replace the IO Acquisition Module*

The typical application of thermal power plant is the application of metal temperature group. Metal temperature of the main steam turbine and boiler and the main auxiliary groups including boiler tube wall temperature, wall temperature of steam turbine, a generator coil and iron core temperature and main auxiliary equipment (such as pumps and fans) temperature. These measured parameters usually changes slowly, does not participate in the system regulation, the remote IO of on-site temperature acquisition and digital processing, then transmitted signals to DCS through the communication bus (also have a way of transmitted signals through control cable to DCS).

In this application, remote I/O is only as a data acquisition system, not through the DCS system on the remote intelligent I/O online configuration. Data are transmitted to the DCS system simply from the remote I/O system [1].

2.2 *Using the Remote IO Station to Form Medium and Small Scale Production Monitoring and Management Information System*

In thermal power plant, some auxiliary equipments part away from the unit, such as water supply and drainage treatment pump station, the exception of ash slag station, heat metering station and booster stations, due to their distance from the main control room far away, so the control is difficult to incorporate into conventional DCS, before we used the independent monitoring system. When the remote I/O is used, the monitoring information of the equipment can be transmitted to the main control room, centralized and unified management.

In this application, remote I/O is not only used as a data acquisition system, but also need to realize the control function. In this case, the communication between the DCS and the remote I/O is a two-way street. The normal operation of a large number of data in remote intelligent I/O and DCS system exchange, at the same time, through the DCS system engineer station can also be configured for remote intelligent I/O [2].

3 Feasibility Analysis of Remote I/O Using in Nuclear Power Plant

The application of remote I/O in nuclear power plant can be considered from the environmental constraints of plant and system, function requirements (operational protection requirements), system performance requirements, installation requirements, equipment identification, cost, other system interface and other point of view, do feasibility analysis of the remote I/O technology application in the nuclear power plant.

3.1 Environmental Restrictions

According to the layout nuclear power plant can be divided into nuclear reactor plant, nuclear auxiliary building, nuclear fuel plant, steam turbine plant, electrical plant and BOP building. Different areas of the application conditions of requirements, Application of remote I/O system should meet the RCC-E (2005) d 2100 provisions of regional environmental restrictions, such as temperature, humidity, pressure, radiation, including:

- Environmental conditions when the device is not running.
- External safety shell normal environmental conditions and accident condition.
- Normal environment conditions and accident conditions inside the safety shell

Some areas have certain radiation, so it is necessary to carry on the radiation supplementary test to the equipment used in this area [3].

Consider from environmental constraints, for the reactor building, nuclear auxiliary plant, nuclear fuel plant, because of radiation and other environmental problems, the remote I/O should not be used. For the turbine plant, electrical plant and BOP building, environmental conditions can meet the basic requirements of remote I/O, you can consider the use of remote I/O system.

3.2 Operation and Protection Requirements

Based on the importance of the executive function, distributed signal transmission requirements and smart device application status, the principle of following system or equipment should adopt the conventional DCS (hard wired connection), should not be application of field bus remote I/O:

- The system has a significant impact on the safety of the reactor and turbine engine (such as RPR, GSE etc.).
- Important control circuit, such as NSSS, GRE, etc.

Consider the system operation and protection requirements, Nuclear reactor plant, nuclear auxiliary plant, nuclear fuel plant and electrical plant ventilation system are not suitable to use the remote I/O system, because the bus signal may be disturbed. Conventional island auxiliary system and BOP plant can be considered using remote I/O system.

3.3 System Performance

The precision and response time of the system (such as SOE) should not be based on remote IO system.

Field bus to meet the main performance indicators of the protection system (response time: 150 ms, sampling accuracy of 0.25%), So in addition to the reactor building, nuclear auxiliary building, nuclear fuel plant, steam turbine plant, electrical plant and BOP building on the accuracy and response time without the special requirements of the system basically meet the system performance requirements by remote I/O Technology.

3.4 Installation Requirements

Should follow the DL/T 1212 on the field bus equipment, communication cable/optical fiber and grounding installation technical requirements, such as plant restrictions or site construction conditions do not meet the requirements of DL/T 1212, remote I/O should not be used [4].

3.5 Equipment Identification

Select field bus system and equipment should be according to the importance of the safety function of the equipment, and carry out corresponding equipment identification tests, identification steps and procedures see volume requirements RCC-E B.

For safety level systems, the control system is required to carry out relevant identification, According to the research of the remote I/O system on the markets, the current remote I/O device can't meet the requirements.

Therefore, only the non safety level of the plant or system using remote I/O.

3.6 Cost

Some system and equipment which are not far away from DCS electronic equipment room, can be not consider using a remote I/O, due to the adoption of the remote I/O do not lower investment cost. Some sub item far away from the DCS electronic devices room, using the intelligent remote I/O technology can reflect the advantages of remote I/O.

From the cost considerations, only some BOP building distance from DCS electronic equipment room, can reflect the cost advantage of remote I/O.

4 Feasibility Analysis of Nuclear Power Plants Using Remote I/O

According to the second section, for the reactor building, nuclear auxiliary building, nuclear fuel building, consider to environmental conditions (such as radiation), operational protection system requirements, and the distance between the DCS electronic equipment room, without consider using the remote I/O system.

For the steam turbine building, due to the short distance between the DCS electronic equipment room, operation and protection requirements, system performance and other factors, in principle, it is not considered the use of remote I/O.

Only some BOP sub term which distributed in the main plant around, there is a need to use remote I/O.

There is no radiation in most BOP building of the nuclear plant, the environmental conditions between the nuclear power plant and the thermal power plant is no obvious difference, except the individual execution level of nuclear safety and nuclear safety related functions of BOP building, plant environment meet the basic conditions of remote I/O equipment.

4.1 BOP Buildings Controlled by Local PLC

The hydrogen station, air compressor room, condensed water fine treatment plant and other BOP buildings and systems are relatively independent. In design of nuclear power plant, by the local PLC achieve local control, the control cabinet is arranged in the BOP buildings. For this type of equipment and systems are relatively independent sub items, the use of remote I/O is not significant [5].

4.2 BOP Buildings Controlled by DCS

BOP buildings controlled by DCS mostly is the combined pump station. In CPR1000 project, most system in combined pump station adopts the traditional DCS, The signals are sent to DCS electric equipment room directly from combined pump station.

Through some experience feedback from the projects under construction and in operation, combined pump station using traditional DCS control have some problems as follows:

- Combined pump station is far from DCS electric equipment room, signal transmission path is long, for some analog signals, appeared significant voltage drop.
- Combined pump station due to the extreme conditions of coast (such as fish signal period, Seaweed flood season etc.), circulating water pump mal operation problem, the corresponding design change more, after the cable tray sealing, appeared with increase measuring points and cables to DCS many times, resulting in site construction difficult, poor scalability, if using remote I/O, the situation can be improved.
- Signals are sent to DCS cabinet through hard wiring, the more cables, high costs. Due to the huge number of cables, the DCS electric equipment rooms always have cable over capacity problem.

Through above analysis, combined pump station has the necessity and feasibility of using remote I/O.

5 Remote I/O Program Implementation of Combined Pump Station

5.1 Remote I/O Program of Combined Pump Station Under the CPR1000 Technical Route

The main process system of combined pump station include: circulating water filtration system (CFI), circulating water system (CRF), circulating water pump lubrication system (CGR), essential service water (SEC), fire water production system (JPP), nuclear plant pump station ventilation system (DWS), circulating water pumping Station ventilation system (DVP), sewage collection system (SEO), etc. CFI/CRF/CGR/SEC/JPP/DWS are taken into DCS control, and CFI/SEO are taken into local PLC control.

In CPR1000 project, circulating water pump station and essential service water pump station jointly build, essential service water system and circulating water system using common filtering. Essential service water system is safety function system, circulating water filtration system and SEC supporting ventilation system DWS are safety related function system. The specific I/O points are shown in Table 1.

In CPR1000 project combined pump station did not using remote I/O. The main reason is as follows.

1. There is no instrument control equipment room in CPR1000 project of combined pump station, instrument control equipment to be shared with the electrical equipment in electrical room. Electrical room congestion has poor environmental conditions. Electrical room layout is as shown in Fig. 1.

Table 1 Specific I/O measuring points of CPR1000 project pumping station

Sys.	Total	Single unit							
		1E				NC			
		DI	DO	AI	AO	DI	DO	AI	AO
CFI	337	86	24	–	–	62 99	12 50	4	–
CGR	46	–	–	–	–	29	16	1	–
CRF	134	–	–	–	–	64	30	40	–
SEC	196	20	12	6	–	82	–	54	22
DWS	62	16	8	–	–	26	12	–	–
JPP	45	6	4	–	–	14	4	17	–

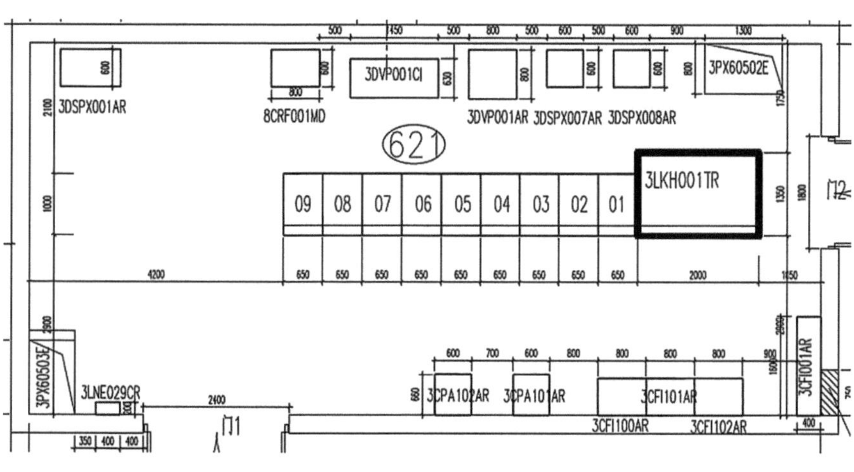

Fig. 1 Combined pump station electrical room layout

2. SEC, CFI and other systems belong to safety and safety related function systems, need to be put in safety DCS to achieve control function, there's no safety remote I/O. This part is not recommended to achieve remote I/O product now.
3. Only more than 200 NC level I/O signals can using the remote I/O, if using remote I/O, different channel of equipment need distribution in different cabinets and remote I/O requires at least three cabinets, and each cabinet arranged 60 measuring points or so, then using the remote I/O not only reduce the cost, there may cause cost increase.

5.2 Remote I/O Program of Combined Pump Station Under the AP1000 Technical Route

Combined pump station provides cooling and production with seawater respectively to circulating water system (CWS), plant water system (SWS), open cooling water system (OWS), circulating water dosing system (WIS), seawater desalination system (SWD) and the drum washing water system.

In AP1000 project combined pump station includes the following process: circulating water system (CWS), circulating water filtration system (WFS), gear box lubricating oil system (GOS), plant water system (SWS), follow the pump room ventilation system (VPS), production wastewater system (WWS), life water distribution system (DPS), plant drainage system (DRS), etc. The number of DCS I/O points is shown in Table 2. Other systems are controlled by local control box.

In AP1000 project combined pump station can use remote I/O. The reasons are as follows:

1. AP1000 as a new designing, can be considered setting independent instrument and control electronic devices room in combined pump station.
2. In AP1000 project, combined pump station does not perform safety related functions, all systems for non-safety system.
3. Combined pump station is far away from DCS electric equipment room, more interlocking with other systems, need to be sent to DCS system.

Table 2 Specific I/O measuring points of AP1000 project combined pump station

System number	Total	Single unit					
		DI	DO	AI	RTD	AO	Communication
WFS	433	337	66	30	–	–	–
GOS	69	33	12	21	3	–	–
CWS	243	78	18	15	90	–	42
SWS	4	The main control equipment in the steam turbine plant, pumping station only 4 AI					

4. The pump station has more measuring points, after using the remote I/O, the advantages of remote IO technology can be fully reflected.

5.3 Remote I/O Program of Combined Pump Station Under the EPR Technical Route

Under EPR project, the pumping station, in addition to the number of flow channels differ from CPR1000, Other are the same as CPR1000 design. Therefore, under the EPR technology route the pumping station is not suitable for the use of remote I/O technology.

5.4 Remote I/O Program of Combined Pump Station Under the HPR1000 Technical Route

Under HPR1000 technical route, circulating water pump station and essential service water pump station have completely independent. SEC system increases the filter section on the original basis, CFI system no longer supply water to SEC system. Circulating water pump station no longer perform safety related functions, including only circulating water filtration system (CFI) and circulating water system (CRF) and circulating pump lubrication system (CGR), circulating water pump station ventilation system (DVP), sewage collection system (SEO) system.

Essential service water system (SEC) and its related supporting system are arranged in essential service water pump station. Essential service water pump station due to the implementation of safety and safety related functions, are still using the traditional DCS control. I/O points of systems in Circulating water pump station with a slight increase compared CPR1000.

HPR1000 project considering use of remote I/O program, the main reasons are as follows:

1. Projects under the HPR1000 technical route are all new projects, the setting and the size of the electronic equipment room, the environmental requirements and so on in the planning stage can be considered completely;
2. HPR1000 project has been confirmed that circulating water pump station and essential service water pump station are separated, circulating water pump station equipment are non safety equipment;
3. The pump station has more measuring points, after using the remote intelligent IO, the advantages of remote IO technology can be fully reflected.

6 Conclusions

As mentioned above, based on the characteristics of the remote intelligent I/O Technology, In this paper, from the nuclear power plant and system of environmental constraints, functional requirements (protection operation, system performance requirements, installation requirements, equipment identification, cost and other system interface and other factors to the feasibility of the nuclear power plant using remote I/O technology are analyzed in detail, determine the scope of the implementation of remote I/O technology for nuclear power plant. Namely in the CPR1000 project and EPR project, do not consider the use of remote IO Technology. In the AP1000 project and HPR1000 project, can be considered using remote intelligent I/O Technology in the pumping station (circulating pump room), to reduce number of cables, enhance the reliability of signals, improve the scalability of control systems in pumping station, and improve the cable capacitance in DCS electronic equipment room.

References

1. Su L S, et al (2007) System and equipment of 900 MW pressurized water reactor nuclear power plant [M] Beijing: Atomic Energy Press, 488–497.
2. Sun S R (2001) Application of remote intelligent IO in the computer monitoring system of thermal power plant [J]. Industrial control computer, 16 (6): 41–43.
3. The French nuclear design and construction rules of association (2005) RCC-E-2005 nuclear electrical equipment design and construction rules of [S].
4. Technical guidelines for the installation of 1212 (2013) DL/T thermal power plant fieldbus equipment [S].
5. GB/T 13286 (2008) Nuclear power plant safety grade electrical equipment and circuit independence criteria [S].

Research on Wireless Location Technology of Nuclear Power Plant and Discussion for Physical Protection System Application

Lin Ye, Hua-Ping Chen, Shuang Li and Wen-Fei Wu

Abstract Several kinds of indoor wireless location technology are introduced in this paper, such as the Radio Frequency Identification (RFID), Bluetooth, Light Emitting Diode (LED) visible light location, and Ultra Wideband (UWB), which are analyzed and compared. Moreover, the paper also discusses the nuclear facilities physical protection system application of wireless location technology in nuclear power plant by means of collecting the location information of the internal authorized personnel to analyze the internal authorized personnel's behavior furthermore for the analysis of the internal threats with the data. And by using the wireless localization technologies it is expected to manage and control the nuclear materials in the real-time and achieve nuclear materials specified in the security area to prevent nuclear materials from being stolen or illegal transfer.

Keywords Indoor location technology · LED visible light location · UWB · Nuclear power plant authorized personnel position · Nuclear material position

1 Introduction

The access control system of Nuclear power plant is only defined in the authorized persons in the area to defend the border through the entry checkpoint, can only be reflected in the access record of the area and the number of personnel in the area, so the situation and the real-time location of the authorized personnel in the area can't be control and nuclear materials temporary control is mainly by means of intrusion prevention technology combined with the video. With the development of power plant intelligence and meticulous management requirements as well as wireless location technology, the location technology is widely used in coal mines and prisons and has good application effect, so we can explore the application of

L. Ye (✉) · H.-P. Chen · S. Li · W.-F. Wu
State Key Laboratory of Nuclear Power Safety Monitoring Technology and Equipment, China Nuclear Power Engineering Co., Ltd., Shenzhen, China
e-mail: yelin@cgnpc.com.cn

Y. Xu (ed.), *Nuclear Power Plants: Innovative Technologies for Instrumentation and Control Systems*, Lecture Notes in Electrical Engineering 400,
DOI 10.1007/978-981-10-3361-2_10

wireless location technology in the physical protection system of nuclear power plant to further achieve the upgrading and improvement of physical protection system of nuclear power plant.

Wireless location technology can't produce electromagnetic interference for the power plant instrumentation and DCS system, so the application information security of wireless network in the security system should be considered.

2 Introduction of Several Kinds of Indoor Wireless Location Technologies

2.1 RFID Position Technology

RFID radio frequency identification technology is a kind of wireless communication technology and also a kind of non-contact automatic identification technology which automatically identifies the target object and obtains the relevant data through the radio-frequency signal. The characteristics of RFID Technology: operation process are non-contact, by the RFID reader and special RFID tag which can be attached to the target object, the information is transmitted from the RFID tag to the RFID reader; RFID tag has the advantages of small volume, low price, easy to control, simple and practical and is currently widely used in the positioning technology. But RFID is subject to the limit of the distance, has a short operating distance and is vulnerable to be interference [1, 2].

2.2 Bluetooth Position Technology

Bluetooth locates by means of measuring the signal strength. This is a kind of wireless transmission technology with a short distance and low-power. The Bluetooth device is small, easy to be integrated and popularized and is easy to be transmitted and not affected by the range of visibility. Bluetooth system is not stable in a complex space environment and greatly interfered by noise and signal.

2.3 UWB Position Technology

UWB is an ultra-wideband communication technology. UWB signal contains a larger range of frequencies than that of narrowband and broadband signals and it is located in the range of ultra-wideband signals. UWB is a new technology and different from other traditional communication technologies. The biggest difference compared with other existing communication technology is that it doesn't need carrier and sends and receives nanosecond extremely narrow pulse data transmission, so the power of signal required in the transmission is low and has a bandwidth with GHz. UWB technology ensures high-speed data transmission while solves the problem of power consumption of the mobile terminal, so there is a wide application prospect. This technology has the characteristics of small transmission power, strong penetrating power, low power consumption, high security, strong anti-interference ability, low system complexity, providing accurate positioning and others [1, 2].

2.4 LED Visible Light Position Technology

LED visible light indoor positioning is a new type of indoor positioning technology, which is the a new wireless communication technology application combined lighting with light communication technology. The technology uses a wide coverage of the LED lighting equipment, can effectively meet the needs of high precision positioning and has low power consumption, strong anti-interference capability, environmental protection and other features. But the localization tags cannot be blocked.

3 Indoor Wireless Location Technology Applications in Nuclear Power Plants

For the transmission rate of UWB positioning technology with ultra wideband technology, higher precision, better stability, stronger anti-interference ability, UWB has a wide application prospect.

LED visible light positioning technology has advantages of low power consumption, high efficiency, small electromagnetic interference effects and other advantages in recent years, so it becomes a very hot research direction.

This paper mainly introduces the UWB positioning system and the system components and system architecture of LED visible positioning system. Referring to the rating grade of the actual protective division and actual situation of nuclear power plant, deploy different positioning system solutions.

3.1 UWB Position System

3.1.1 System Solution

The tag should be set to the items (such as safety helmet, safety shoes) which an authorized officer must carry in the nuclear power plant, or fixed on the monitored items. The base station uses an omnidirectional antenna with the placement of 30 m interval to ensure that the protective area of nuclear power plants is wholly covered.

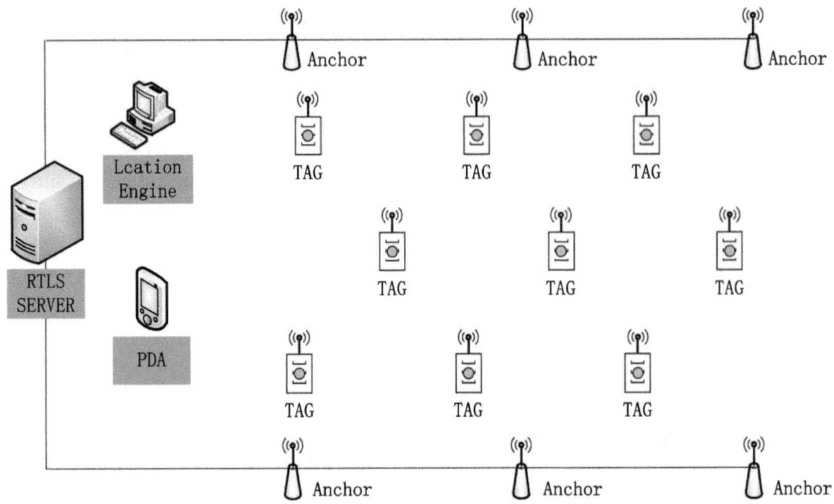

Fig. 1 UWB position net approach

The positioning tag is equipped with the random code sequence. Sensor stations send the pulse signal according to the time set. When it detects a random code sequence, it compares the original code with the random code sequence to achieve precise positioning by using TDOA mechanism. UWB position net approach, see Fig. 1.

3.1.2 System Composition

1. UWB Anchor calculates the distance from TAG and pass back to the server via a wired or wireless network.
2. UWB TAG broadcast their position according to the system requirements.
3. Server.

Position calculation engine: calculate the specific location of the label according to the ranging results between the Anchor and Tag.

Graphic rendering server: Graphics Rendering Server: Tag position is presented in a graphically way based on vector diagram of scene.

Other servers: according to the system application design other servers, such as server for data mining.

3.2 Solution of LED Visible Position System

LED optical communication positioning technology is a communication technology based on visible light, reads the characteristics of the image information by means

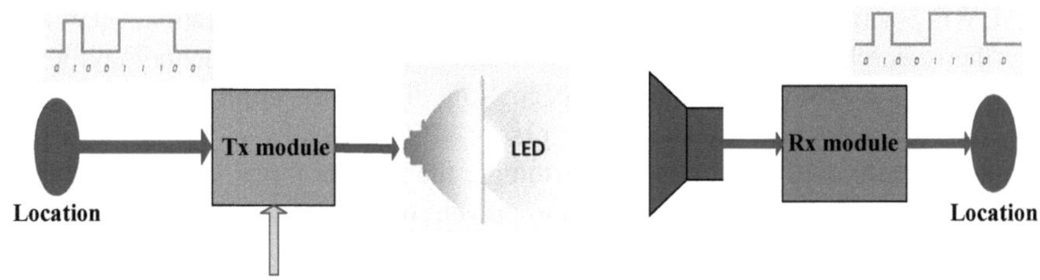

Fig. 2 LED visible light positioning principle

of the image sensor line by line and is modulated by the driving current or driving voltage of the light source, i.e. by changing the frequency and duration in the driving current or the voltage to form different light and dark stripes in the image on the image sensor output. Then according to the detected light or dark stripes' number or phase of image, through the calculation, obtain the frequency information of the driving current loaded. And then determine the code according to the transmitted frequency information is coded. Finally, recover the transmitted information based on the specific decoding mode. In case of practical application through the LED lamps on the roof, lamps emit flashing signals like Moss telegraph code, which are received and detected by the user's smart device or specific equipment and the user can receive direct feedback over the light signal without directing the receiving device at a particular direction. The positioning accuracy may be within 1 m. LED visible light positioning principle, see Fig. 2.

4 Discussion on Wireless Location Technology in the Physical Protection Application of Nuclear Power Plants

There are a large number of electronic devices associated with nuclear safety in the electrical plants of nuclear power island and the barriers of nuclear island reinforced concrete structure, so we should select a smaller electromagnetic interference, stronger penetration of wireless location technologies, the wireless positioning technology furthermore need have the simple and convenient layout features so as not to affect the implementation of major systems of nuclear power plants safety equipment.

At present the access control system of nuclear power plant is only managed by the authorization in the board of security area. It can be only reflected in the state of personnel "in" and "out" of the board and cannot grasp the real time position of internal staff, especially after entering the vital parts. Using the wireless location technology enables personnel in the real-time monitoring and is conducive to radiation safety and security of nuclear materials.

UWB positioning technology and LED visible light wireless location technology are the research focuses of wireless communication positioning technology; in the future it has a broad application prospect. Now discuss the application of wireless location technology in nuclear power plant (NPP) especially in the physical protection system: used for auxiliary determining the internal threat, actively detecting and according with intrusion detection to prevent nuclear material from being stole; other emergency applications in the nuclear power plant.

4.1 Auxiliary Discriminating Internal Threats

Usually when recognizing internal threat of physical protection system in the nuclear power plant is greatly difficult, the positioning technology can be used in the real-time and on-line monitoring for any authorized officer into the protective area of nuclear power plant, collecting and analyzing a large amount of data for any authorized personnel behavior, especially those into the vital parts of nuclear power plant, such as staying for a long time, frequently access to vital parts, together with the other means of identification and management system to effectively reduce the determination range of insider threat of the physical protection of nuclear power plant and provide technologically the auxiliary method for determining internal threat.

4.2 Anti-nuclear Material Missing

The positioning technology can be applied in the staff and also store items. Usually store items are stationary. The receiving device of positioning is fixed on the article which can be monitored in real time whether there has been displacement. For the physical protection of nuclear material, positioning technology makes the state of article from passive defense and anti-invasion to the active discovery, which improves the effectiveness of the physical protection of the three elements of the "detection" feature. It can be expected that the positioning technology in the system of physical protection application of nuclear power plants the feasible.

4.3 Other Applications

When the nuclear power plant is in the state of emergency, the real-time location of staff can assist to determine the location of staff trapped, so that the emergency and rescue are more quick and effective. In the emergency collection region, realize the quickly check and timely evacuation of emergency personnel.

4.4 Difficulties

Compared to wired technology, the key issue of wireless technology applications of nuclear power plant needs to be focused on resolving electromagnetic interference and electromagnetic compatibility. In nuclear power plants there are more mechanical and electrical equipment, the sources of electromagnetic interference abound, which may affect voice communications and operational management based on wireless communications. More importantly, the electromagnetic wave wireless device transmits may cause the interference for critical electrical and mechanical equipment of nuclear power plants and there is a risk of normal reading for instruments.

The positioning label carrier also is the key technology of nuclear power plants and the locating labels should be easy to use for the plant personnel into the plant under the condition where the locating tag should not affect personnel's work in the nuclear power plant.

5 Conclusions

In this paper, through the analysis of wireless location technology, particularly UWB and LED visible light communication technology exposition, with the continuous development of wireless technology, the key problem of wireless location technology and wireless real-time location and risk of electromagnetic interference are solved. Select the positioning technology applied to nuclear power plant in different security area to ensure the operation of nuclear power plant safety; solve problems about the locating label carrier; solve the problem of information security, and potential risk for wireless hardware for the security of the system. The wireless location technology will play an important role in the application of physical protection system of nuclear power plant and daily management and provided a strong technical support for the physical protection of nuclear power plants.

At this stage the wireless location technology may be attempted to be applied in some auxiliary buildings which are not involved in the nuclear safety, which accumulates experience for the comprehensive application of nuclear power plant.

References

1. Chun-Ling TANG, Ming-Gang SUN (2014) Research on personnel location system of mine based on wireless location technology. Coal technology 1:116–118
2. Fan ZHANG, Dian-Cheng CHEN, Jie YANG (2012) Comparative study on indoor location technology and system. Guangdong Communication Technology. 11: 73–79

Analyzing the Digitally Levelled Control Systems Used for Nuclear Power Plants

Yuan-Long Wang

Abstract This paper analyzes the digitally levelled control systems used for the nuclear power plants. The main contents are as follows: Through describing the expression of DCS, the accurate concept of the digitally levelled control system is highlighted. The level model of the digitally levelled control system is provided here by examples. The comparison is done here, in which the difference of classification to the same subsystems in the different digitally levelled control systems used for the nuclear power plants is clearly shown. The digitally levelled control system used for the nuclear power plant is a network system. Its safety belongs to one part of the whole nuclear safety. Here, the paper will concisely discuss the safety problem according to the nuclear safety principle of defense in depth. But the notice must be pointed out that the principle of defense in depth does not only exist in the nuclear energy systems, also in the industrial control systems.

Keywords Nuclear power plant · Digital · Level · Control system · Function safety

1 Introduction

At China, the widespread adoption of DCS technique to configuring the new instrumentation and the control systems in nuclear power plants under construction has been already practiced [1, 2].

Here, DCS can be in the name of Digital Control System, or Distributed Control System. The following description will show that doing distinguishing between them is necessary.

Figure 1 is the schematic diagram of the digital control system [3–6]. Here, the actual meaning of word 'digital' is that computer or microprocessor is widely used in the instrumentation systems.

Y.-L. Wang (✉)
Science and Technology Laboratory on Reactor System Design,
Sichuan, China
e-mail: wang_yuanlong126@126.com

Y. Xu (ed.), *Nuclear Power Plants: Innovative Technologies for Instrumentation and Control Systems*, Lecture Notes in Electrical Engineering 400,
DOI 10.1007/978-981-10-3361-2_11

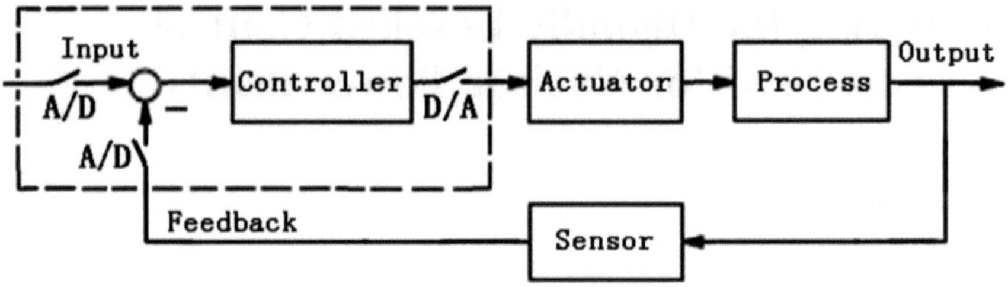

Fig. 1 The schematic diagram of the digital control system

In the digital control system, A/D converter can convert the analog signal (input) into the digital signal. But the actuator needs the analog signal, so the digital signal must be converted into the analogy signal through D/A converter.

The distributed control system has the feature with levels or layers. Its theoretical basis is built on the levelled control of a large system [7]. The control of the large system is relied on the control systems in it. Such control systems cooperate with each other to implement the control function of the large system.

Distributed means hierarchy, i.e., levels. So, the distributed control system is called the levelled control system. Figure 2 shows the level classification of the levelled control system [8, 9].

In Fig. 2, the information communication among levels is the basis of the distributed control system. So, the distributed control system is also called the information control system.

It is difficult for the analog control system to implement the information control based on levels. To implement the information control, the prerequisite is the application of computers or microprocessors. Through integrating the automatic control and the controlled object as a unit, a digital control and supervisory system is constructed. As the enterprise management and process control are classified as levels, a hierarchy model is built effectively. Therefore, the digitally levelled control

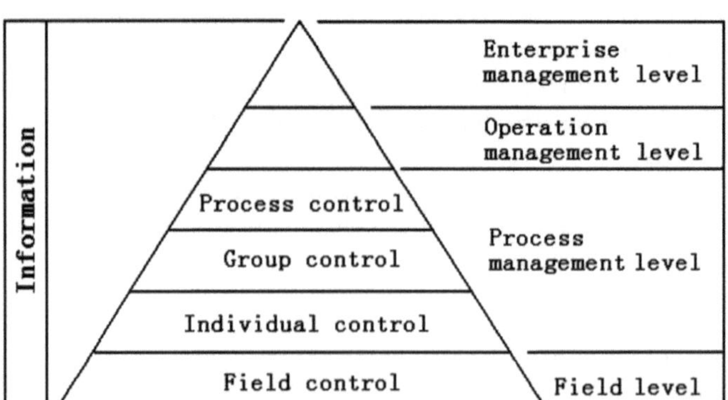

Fig. 2 The control levels of a large system

system of a large system has been completed. Here, it is obvious that the current DCS should be named as the digitally leveled control system, i.e., DLCS. The DCS is only used for the digital control system.

2 Typical Nuclear Power Plant and Control Systems

The schematic diagram of the typical nuclear power plant (NPP) is shown at Fig. 3 [10]. Normally, a nuclear power plant is classified into two loops. The primary loop is the nuclear island. The secondary loop is the normal island. Their main devices are highlighted in Fig. 3. To secure the nuclear power plant, all of these devices must be controlled and monitored.

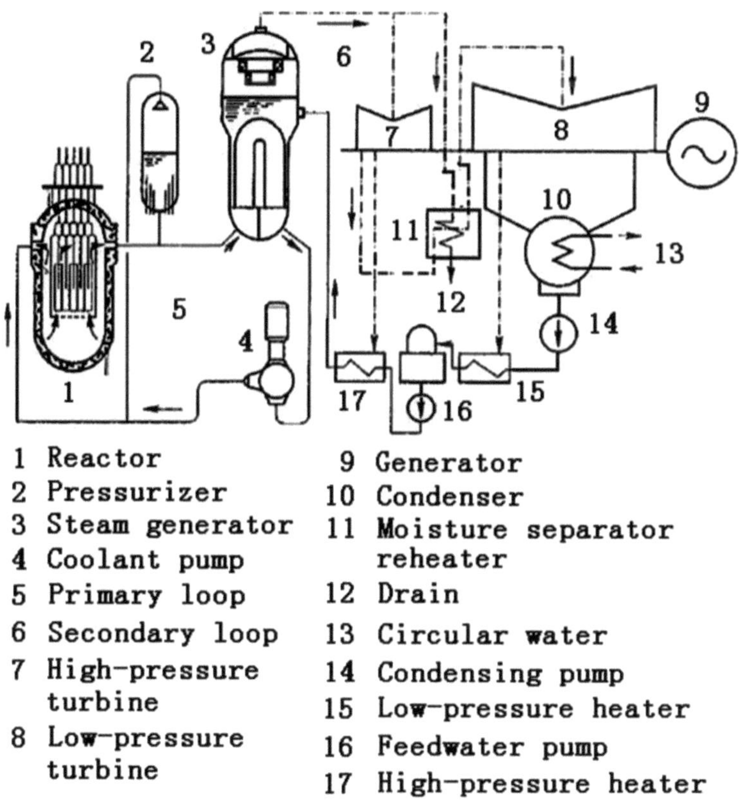

Fig. 3 The schematic diagram of typical nuclear power plant

2.1 The System Scale and the Control Scale of A Typical Nuclear Power Plant

The fact of the history of NPP is clear that its thermal-hydraulic loop number, for example of 1000 MW NPP, has experienced three stages: from four loops to three loops, and then to two loops [11]. Here, if the estimation is done according to the mature model of NPP, i.e., the second generation or the second generation modified, the control signal quantity of the whole system is listed in Tables 1 and 2 [12].

Notice, Tables 1 and 2 are only to Tian-wan NPP in China. The statistic data to other NPPs will be not the same as here. Especially, AP1000 NPP is absolutely different from here [11].

2.2 NSSS Control Systems in NPP

Through Fig. 3, Tables 1 and 2, it is clear that a NPP is a typical large system. Such system needs the levelled control policy.

Here, for analyzing concisely, the example is only to the control systems of the nuclear steam supply system (NSSS) of NPP. Figure 4 is its schematic diagram [10, 13, 14]. In Fig. 4, the control systems include as follows.

The reactor power control system. its function is to control the reactor power (nuclear power). Its measurement inputs are nuclear power, temperatures respectively come from the hot section and the cool section of the coolant of reactor, the load needed by turbine, etc.

The pressurizer pressure and level control system. The inputs are the pressure and level of the pressurizer.

The steam generator (SG) level control system. The inputs are the steam flow, the feedwater flow, the load needed by turbine.

The steam dump control system. The inputs are the load needed by turbine, the average temperature of coolant.

Table 1 1000 MW NPP control signals

Type	Temperature	Pressure, level, flow	Electrical measurement	On-off input	On-off output	Analog outpour	Local indicator
Number	2500	3100	450	900	500	330	1000

Table 2 Actuators and control loops

Type	Motor	Isolated value with switch	Isolated value without switch	Control value	Solenoid value	Heater	Closed control loop
Number	1000	2850	1700	350	120	300	330

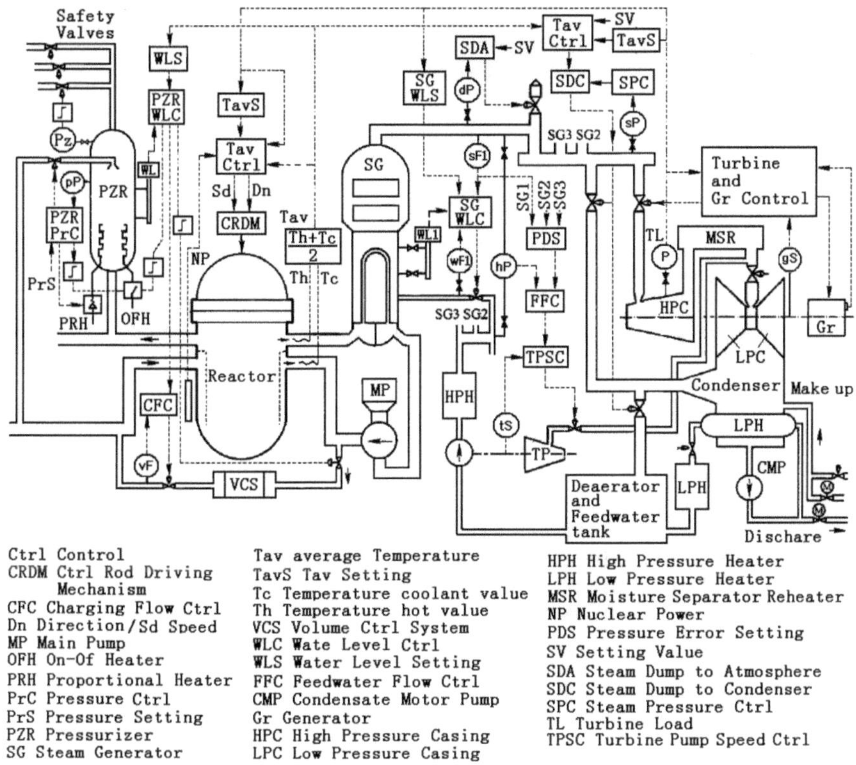

Fig. 4 NSSS control systems in NPP

3 The Model of the Digitally Levelled Control System of NPP

According to the levelled policy of Fig. 2, the digitally levelled control system is shown at Fig. 5 [8, 9]. Here, the control system consists of four levels. In Fig. 5, the rest control systems of NSSS excluding the steam dump control system belong to SICS.

Figure 6 is the schematic diagram when the levelled architecture of Fig. 5 is replaced by the actual engineering DCS [8, 9]. The classification of levels in Fig. 6 corresponds to the levels from Field level to Process management level in Fig. 2. The meanings of abbreviations in Fig. 6 will be introduced in Fig. 7.

Here, as an example of NSSS control systems, Fig. 6 is the exhibition of the levelled policy of all the process control systems except the reactor power control system. The process control systems are the non-safety to the nuclear safety consideration. Their instrumentation configuration is accomplished by TXP automatic control system in Siemens. To the safety system and the relevant-safety control system, for example, the protection system and the reactor power control system,

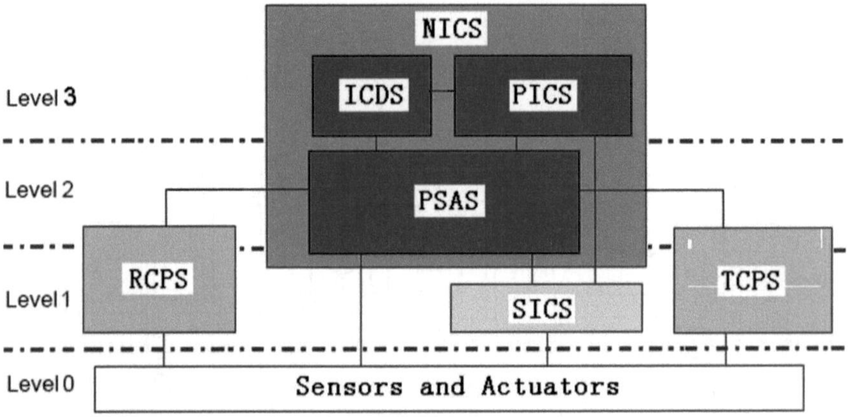

IC Instrumentation and Control
ICDS IC Configurating and Diagnosis System
NICS Normal IC System
PICS Process Information and Control System
PSAS Plant Standard Automation System
RCPS Reactor Control and Protection System
SICS Special IC System
TCPS Turbine Control and Protection System

Fig. 5 Digitally levelled control system (1)

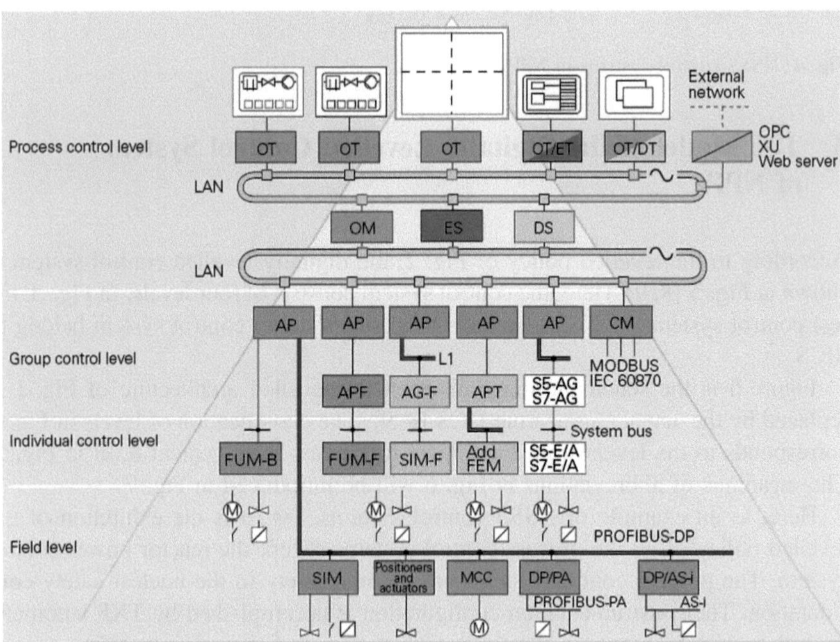

Fig. 6 Digitally levelled control system (2)

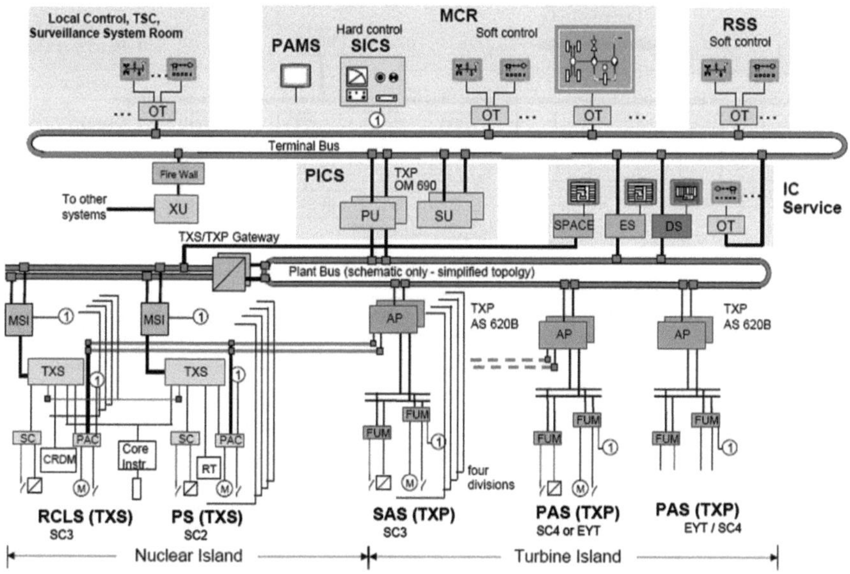

AP Automation Processor
DS Diagnosis System
ES Engineering System
IC Instrumentaion and Control
OM Operating and Monitoring
OT Operating Terminal
PS Reactor Protection System
PU Data Processing Unit
XU Connecting Unit
SC Safety Cabinet
SU Data Storage Unit
FUM Function Unit Module
MCR Main Control Room
MSI Monitoring Service Interface
PAC Process Automation Cabinet
PAS Process Automation System
RSS Remote Shutdown System
SAS Safety Automation System
TXP Normal IC System
TXS Safety IC System
CRDM Control Rod Driving Mechanism
PAMS Post Accident Monitoring System
PICS Process Instrumentaion Control System
RCLS Reactor Control and Limiting System
SICS Safety Instrumentaion Control System
SPACE Specification and Coding Environment

Fig. 7 Digitally levelled control system (3)

Siemens uses TXS automatic control system to implement. The digital instrumentation and control systems of the whole NPP are concisely exhibited by Fig. 7 [9, 12].

Here, the notice must be taken into consideration, i.e., in Fig. 7, both the reactor power control system(RCLS) and the reactor protection system (PS) are implemented by TXS automatic control systems, and Westinghouse, however, considers the reactor power control system (RCS) as being non-safety (see Fig. 8). It is obvious that the different considerations exist when the actual control systems are replaced by the digital instrumentation systems [15].

By comparing Fig. 8 with Fig. 6, the other differences will be exposed. First, the word phrase meaning for each level is not the same. It reflects the fact that the design thinking on how to construct the digitally levelled control system is not the equivalent. Second, the main network shows the similar problem. Figure 6 uses the circular net, and Fig. 8 uses the bus net [16]. If all these imply that the advantages and the disadvantages exist each other. The problems are worth to study further.

Here, because the pages are limited, the management levels in Fig. 2 do not be discussed [17, 18].

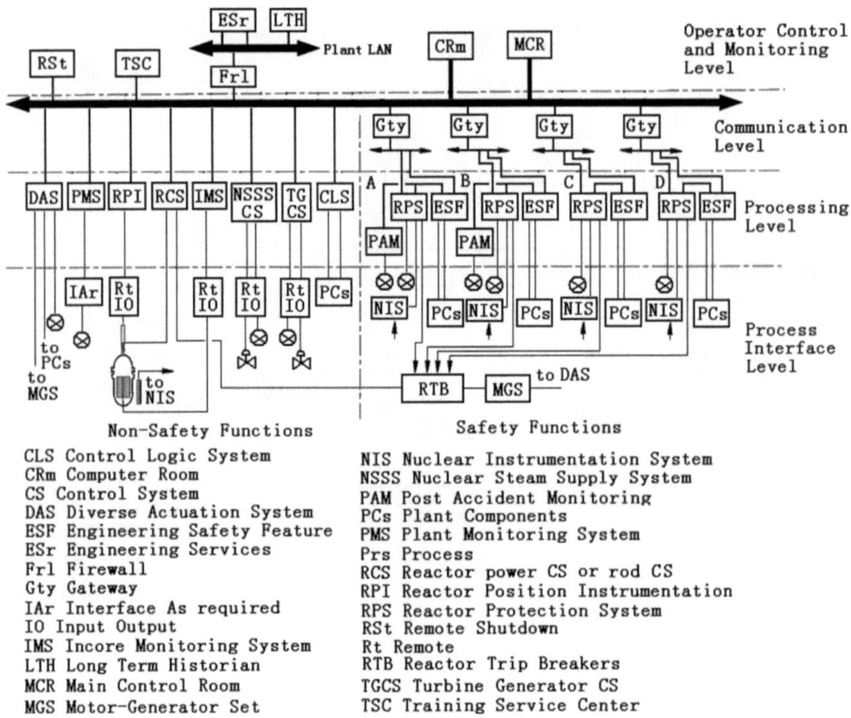

Fig. 8 Digitally levelled control system (4)

4 The Safety Problem of the Digitally Levelled Control System

The radioactive problem which especially exists in nuclear energy systems causes the safety problem on the application of the nuclear energy. According to the nuclear safety principle of defense in depth [19], the reactor control systems belong to the first safety layer (level).

In the view of the above nuclear safety principle, NPP is classified into two loops: the primary loop and the secondary loop (seeing Fig. 3). Here, the primary loop corresponds to the nuclear island in Fig. 7, and the secondary loop corresponds to the turbine island in Fig. 7. Based on this classification, the digitally levelled control system is also divided into two parts: for the safety functions and for the non-safety functions (seeing Fig. 8). To five control systems above, the reactor power control system (RCS) is the relevant safety control system. The others (NSSS CS) are the non-safety control systems [20]. But here, the notice must be taken into consideration, i.e., in Fig. 8, RCS and NSSS CS are all considered as the non-safety control systems. And in Fig. 7, RCS is the part of RCLS in the nuclear island. The difference which exists here will affect selecting the devices of the individual control system.

Even though the safety classification is not the same here, the four-level classification of the digitally levelled control system is not affected. Here, another notice must be pointed out is that the principle of defense in depth not only exists in the nuclear energy systems, but also exists in the industrial control systems [18]. The details are not discussed further.

5 Conclusions

Here, the principle analysis of the digitally levelled control system used for NPP has been finished. The architecture for discussion comes from the Siemens products which have been applied in Tian-wan NPP.

In the Chinese mainland, the public view has been realized in the nuclear industry, that is, using the digital technology to build the new nuclear power plants is the intelligent selection. Because of the factor of the history, the current situation is that the digitally levelled control systems used for the nuclear power plants are mainly introduced abroad, and the way is not only one. The introductions from the more suppliers easily produce the problem that it is difficult for users to unite the technology model. So it will result in the effect that the costs for construction, operation, and maintenance will obviously increase. In order to avoid the phenomenon appears repeatedly, it is important for the users to enhance the force to study the principles on the digitally levelled control system and develop the relevant techniques. It is obvious that the advanced techniques are mastered and then the products could be manufactured and used safely.

References

1. China nuclear society, china instrumentation society (2013) Theses on the second China nuclear power instrumentation and control conference [G]. Xian, China.
2. China nuclear energy industry society (2014). Theses on the digital instrumentation and control system technology conference of the nuclear energy industry [G]. Shanghai, China.
3. Shi Ren, Liu Wen Jiang (2005) Automatic instrumentation and process control [M]. Electronic Industry publisher, China.
4. Shao Yusen (1995) Process control and instrumentation [M]. Shanghai Jiaotong University Press, China.
5. Curtis D. Johnson (2002) Process control instrumentation technology (sixth edition). M Printed by Science Publisher, USA.
6. Wang Yuanlong (2016) Digital controller technology and application. J China Instrumentation, 23: 56–60.
7. Qian Xuesen, Song Jian (1980) Engineering cybernetics [M]. Science Press, China.
8. I&C working team (2004) Digital instrumentation and control platforms Teleperm XS/XP. G Siemens ANP, FRANCE.

9. Siemens PGL, Areva NP NL-G (2006) Fang Jiashan nuclear power project I&C exchange meeting. G Siemens.
10. Guang Dong nuclear power training center (2005) 900 MW PWR nuclear power plant systems and devices. M Atomic Energy Publisher, China.
11. Wang Yuanlong (2013) 1000 MW nuclear power plant nuclear island architecture and safety comparison analysis [C]. China Atomic Energy Publisher, China.
12. Jiang Su nuclear power limited company (2002) Tian Wan nuclear power plant digital instrumentation and control systems. G JSNPLC, China.
13. Cui Zhenhua, Wang Yuanlong (1998) NSSS control systems simulation program research report [R]. Tsinghua university and NPIC cooperating team, China.
14. Liao Zhongyue, Wang Yuanlong (1997) NPIC-CNEIC-QINSHAN-FRAMATONE Cooperation Agreement: Catia2 Code-First part: Descriptions of the models. R NPIC, China.
15. Wang Yuanlong (2013) Nuclear power plant digital instrumentation and control systems architecture comparison analysis [C]. China Atomic Energy Publisher, China.
16. Bnfl (2008) Common Q-Ovation Integrated I&C Platforms. G WESTINGHOUSE, USA.
17. Michael Palmer, Robert Bruce Sinclair (2003) Guide to designing and implementing local and wide area networks (second edition). M Printed by Tsinghua University Press, China.
18. Xiao Jianrong (2015) Industrial control system information safety. M Electronic Industry Publisher, China.
19. International atomic energy agency (2012) Safety standards series No. SSR-2/1 Safety of nuclear power plants: design. S IAEA, Vienna.
20. International electrotechnical commission (2009) IEC 61226 Nuclear power plants-Instrumentation and control important safety-Classification of instrumentation and control functions [S]. IEC, Geneva, Switzerland.

A Study About Software Development QC and QA of the Digital RPS in Nuclear Power Plant

Wei-Hua Chen, Wang Xi, Peng-Fei Gu and Wang-Ping Ye

Abstract Reactor protection system (RPS) is one of the very important systems in nuclear power plant (NPP). Since the digital technology was used in RPS, for the software development,the quality control (QC) and quality assurance (QA) should be further researched. Based on the CPR1000 digital RPS software development project, this paper discusses the definitions and necessity for QC and QA, illustrates the software life cycle, proposes the practical verification and validation (V&V) activities, describes V&V process through in software development. In each stage of V&V process, the V&V activities that QC and QA works and products implemented by are detailed introduced, provides a practical and efficient reference of QC and QA method for other nuclear digital safety system software development projects.

Keywords NPP · RPS · QA · QC · V&V

W.-H. Chen · W. Xi · P.-F. Gu · W.-P. Ye
Instrumentation and Control Department,
China Nuclear Power Design Co., Ltd.,
Shenzhen, China
e-mail: chenweihua@cgnpc.com.cn

W. Xi
e-mail: wang.xi2@cgnpc.com.cn

W.-P. Ye
e-mail: yewangping@cgnpc.com.cn

P.-F. Gu (✉)
Institute of Nuclear and New Energy Technology,
Collaborative Innovation Center of Advanced Nuclear
Energy Technology, Tsinghua University, Beijing, China
e-mail: gupengfei@cgnpc.com.cn

Y. Xu (ed.), *Nuclear Power Plants: Innovative Technologies for Instrumentation and Control Systems*, Lecture Notes in Electrical Engineering 400,
DOI 10.1007/978-981-10-3361-2_12

1 Introduction

RPS is one of the most critical Instrumentation and Control (I&C) systems, which protected the NPP by triggering reactor trip and actuating the engineered safety features. The previous RPS designs were based on analog technologies, which preferred to assure and control the quality of hardware, such as hardware identification and environment test. Compared with traditional analog technology, the application of DCS can effectively improve the accuracy, stability, safety and reliability [1] of the nuclear power plant. Therefore, the quality of RPS software led to widespread concern, and the quality control (QC) and quality assurance (QA) for the software development of digital RPS should be further discussed.

The safety software plays a key role in digital RPS devices, its fault may lead to system operation failure. QA and QC, which were implemented to reduce the defects during software development, have been the focal points to realize the expected functions in RPS. The QA and QC in software development can be implemented by the independence V&V activities, which complying with procedures and standards strictly. To detect the mistakes and provide information for error repairing, V&V activities are executed in the whole stages of software development.

Combining with the V&V activities in CPR1000 safety digital RPS design project, this paper makes a discussion for QC and QA, and introduce the practical V&V activities in software development.

2 Discussions

2.1 Definitions

QC activities including the process checking, quality evaluation, and tested or reviewed product concerning, aims to find out the potential and possibility mistake or error during software development, and proposes information for error repairing.

QA activities implemented complying with procedures and plans, aims to guarantee consistency between software development and expected results, reduces the differences between the actually and expected outcome, It is also obligated to control the operation of QA system, the differences between the implementation and specification, and proposes analysis reports and suggestions during software development.

In software V&V activities, verification conforms to requirements for all activities during each life cycle process, satisfies the standards, practices, and conventions during lifecycle processes, successfully completes each lifecycle activity and satisfies all the criteria for initiating succeeding lifecycle activities. Validation satisfies system requirements allocated to the products at the end lifecycle activity, solves the right problem, and satisfies intended use and user needs in the operational environment [2].

2.2 *Necessity*

QC and QA are emphasized in legislations and standards to insure the safety and reliability of systems in nuclear power plant.

In nuclear safety legislation HAF 102 [3] and its related guide HAD 002/16 [4], the requirements are provided for the application of safety computer system and the design of safety software, including the formulation and implementation of the standards, and the technique method for software QC and QA. The QA outline indicates that the V&V activities are necessary for the software development process. HAF 003 [5] and its related guide HAD 003/06 [6] provides the QA requirement for safety items.

R.G.1.152 [7] indicates that the QA for safety software need QA plan and software V&V activity. The V&V process of the overall system should ensure the testability, correctness, consistency, completeness, and accuracy of the system security requirements.

IEC 60880 Standard for software used in nuclear power plant safety systems [8] and IEEE 1012 Standard for software V&V provides the work, range, method and document-specific for V&V activity.

The structure of legislations and standards for QA and QC, software lifecycle and V&V are shown in Fig. 1.

The description of the legislations and standards illustrate the importance and necessity of V&V activities in nuclear safety software development. The V&V activity complying with the legislations and standards in RPS software development is a basic guarantee for QC and QA.

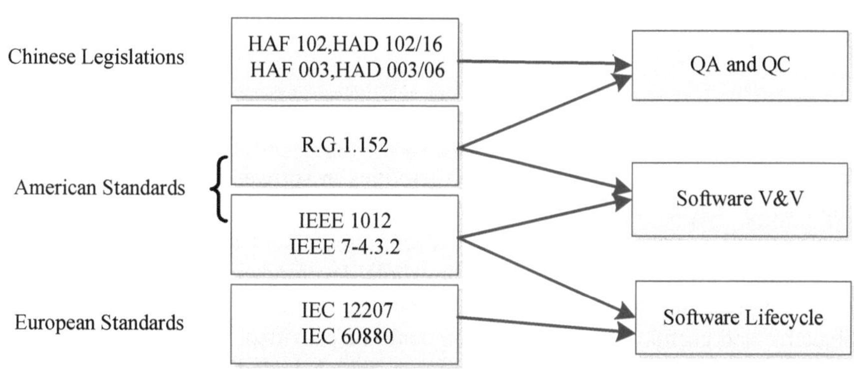

Fig. 1 The structure of legislations and standards

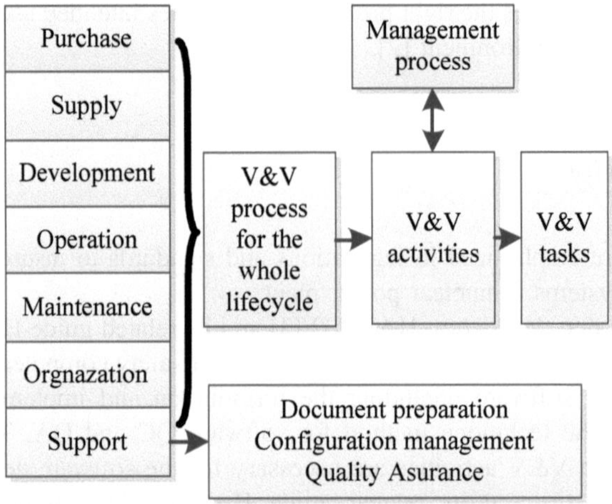

Fig. 2 Software life cycles

2.3 Software Lifecycle

The software lifecycle described in IEC 12207 [9] is shown in Fig. 2, which were referenced in IEEE 7.4.3.2 [10] and IEEE 1012, including purchase, supply, development, operation, maintenance, organization, and support processes. Among them, the development process draws significant attention in CPR1000 safety digital RPS design project. According to IEEE 7.4.3.2 and IEEE 1012, V&V management process, activities and tasks exist in the whole software lifecycle, the V&V activities and tasks corresponding with each process should be complemented.

3 Stage QC and QA in V&V Activities

3.1 V&V Process

Combing with CPR1000 nuclear power project, according to IEC 60880 and IEEE 1012, the V&V process for RPS software design and development has been instructed in Fig. 3, as it shown, V&V activities in software development process contains 6 stages, including concept stage, requirement stage, design stage, implementation stage, and integration stage, all of them are independent of the development process, make tests for the whole development process.

To implement the QC and QA in RPS software development, assure the objective and useful default information could be provided, the V&V activities are carried out by the third party independence, which including the technical independence, managerial independence and financial impedance.

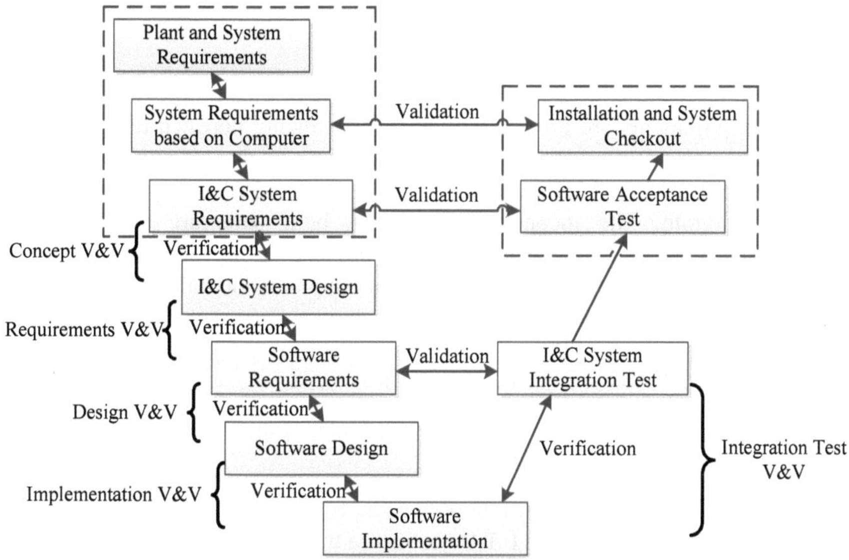

Fig. 3 The V&V process for the RPS software in CPR1000 RPS project

3.2 V&V Plan

At the beginning of V&V project, the V&V plan was provided to describe the purpose, definitions, overview, processes, activities and tasks. To implement QC and QA, the V&V plan should be reviewed and the QA plan should be prepared at the same time.

Meanwhile, according to the legislations and standards, the writing and publication of documents should comply with the specifications. For instance, required by R.G.1.172 [11], in order to provide standard, traceability, and verifiability documents for V&V activities, the developer was obligated to write and publicize the software requirement specification document complying with strict specification.

3.3 Concept V&V

In concept V&V, quality is controlled by requirement tracing, including forward and backward, between system requirement specification and 16 design scheme introductions. The non-conformance terms are reviewed and classified into Concept V&V Bug Report, which is provided to developer for error restore. QA is worked by auditing, including concept V&V process audit and delivery audit, the results are gathered in Concept V&V Report to evaluate concept V&V activities.

3.4 Requirement V&V

During requirement V&V, QC worked by requirement tracing in two parts. In the first part, requirements are forward and backward tracing between design scheme introductions and software requirement specification. In the second part, requirements are forward and backward tracing between Logic/Analogy Diagram and Function Diagram. The inconsistent terms in both parts are emphasised in Requirement V&V Bug Report, which is propose to developers for default repairing. QA works through this stage to audit the process and delivery, and make a conclusion in Requirement V&V Report.

3.5 Design V&V

In design V&V, quality control is implemented by forward and backward tracing between I/O interfaces and EAST TAG to insure the whole interfaces are designed correctly. The mistakes in each interface are illustrated in Design V&V Bug Report, which is provided to developer for interfaces correctly designed. QA still makes process and delivery audit, and produces Design V&V Report.

3.6 Implementation V&V

In implementation V&V, QC executed by test case design, unit test and regression test. In this stage, the test cases are carefully designed for each algorithm block, which is tested in six rounds. After each round, the V&V Bug Report is produced for the error repairing by developer, and the related regression test is executed to evaluate the repairing efficiency. QA executed by evaluating the product quality state evaluation, process audit and implementation delivery audit. Implementation V&V report and Product evaluation report are produced after QA works.

3.7 Integration Test V&V

In integration test V&V, QC contains works for integration test and system test. In these works, the test cases are designed for typical functions, which covered the main unit and algorithm blocks to test their integration performance. The integration V&V Bug report in this stage is produced to guarantee functions and systems operated correctly. Process audit and delivery audit are also implemented to assure the quality, in additional, the Integration V&V report and Product evaluation report are produced to show the QA result.

Table 1 V&V activities for QA and QC in each stage

Category	Quality assurance (QA)		Quality control (QC)	
Stage	Works	Products	Works	Products
V&V plan	QA plan preparation	QA plan	V&V plan management Review	V&V plan
Concept V&V	Process audit Stage delivery audit	Stage V&V report	Requirement tracing Review	V&V bug report
Requirements V&V	Process audit Stage delivery audit	Stage V&V report	Requirement tracing Review	Bug report Test report
Design V&V	Process audit Stage delivery audit	Stage V&V report	Requirement tracing Review	Bug report Test report
Implementation V&V	Product quality state evaluation Process audit Stage delivery audit	Stage V&V report Product evaluation report	Test case design Unit test Regression test Test management Review	Test case Bug report Test report
Integration test V&V	Product quality state evaluation Process audit Stage delivery audit	Stage V&V report Product evaluation report	Integration test System test Test management	Bug report Test report

The works and products of V&V activities for QA and QC in each stage are concluded in Table 1.

To meet the requirement of the software integrity level, considering the different key point of each stage, the different standard should be referenced and introduced. For example, the unit test in implementation V&V stage could comply with the aerospace and military industry standard, which contains skilled experience to meet the requirement of V&V of digital RPS software in nuclear power plant [12].

4 Conclusions

This paper discusses QC and QA that practiced and implemented combining with CPR1000 digital RPS software development project, based on the importance and necessity that emphasized by legislations and standards, QC and QA works and products implemented by the third party independence V&V activities are detailed introduced in each stage of software development, provides a practical and efficient reference of QC and QA method for other nuclear digital safety system software development projects.

Acknowledgements This project was financially supported with funds which provided by a reliability research of reactor protection system which based on digital microprocessor and electrical equipment (national science and technology major project). The award number is 2014ZX06004002-004.

References

1. Liu Z, Hu L S, Bai T (2015) Requirements and Methods of Reliability Design for Safety Level Software of Nuclear Power. PROCESS AUTOMATION INSTRUMENTATION, 36(11):116–120.
2. Software Engineering Standards Committee of the IEEE Computer Society (2004) IEEE 1012 IEEE Standard for Software Verification and Validation. Institute of Electrical and Electronics Engineer, New York.
3. HAF 102 (2004) Safety of Nuclear Power Plant Design Regulations. Doctoral dissertation.
4. HAF 102/16 (2004) Safety of Nuclear Power Plant Design Regulations Guides. Doctoral dissertation.
5. HAF 003 (1991) Safety of Nuclear Power Plant Design Regulations. Doctoral dissertation.
6. HAD 003/06 (1986) Safety of Nuclear Power Plant Design Regulations Guides. Doctoral dissertation.
7. R.G.1.152 (2006) Criteria For Use Of Computers in Safety Systems of Nuclear Power Plants [S]. U.S Nuclear Regulatory Commission.
8. International Electro technical Commission (2006) IEC 60880 Nuclear power plants-Instrumentation and control systems important to safety-Software aspects for computer-based systems performing category a functions. International Electro technical Commission, Switzerland.
9. Joint Technical Committee ISO/IEC JTC 1, Information technology, Subcommittee SC 7, Software engineering (1995) IEC 12207 Software Life Cycle Processes. International Electro technical Commission, Switzerland.
10. Nuclear Power Engineering Committee of the IEEE Power Engineering Committee (2010) IEEE 7-4.3.2 IEEE Standard Criteria for Digital Computers in Safety Systems of Nuclear Power Generating Stations. Institute of Electrical and Electronics Engineers, New York.
11. R.G.1.172 (1997) Software Requirement Specification For Digital Computer Software Used in Safety Systems of Nuclear Power Plants [S]. U.S NUCLEAR REGULATORY COMMISSION.
12. Ding Y X, Gu P F, et al. (2015) Study on Standard about Safety Digital I&C System in NPP. Process Automation Instrumentation, 36 (11):61–64.

A Study of NI Indoor Wireless Coverage in CPR1000 Projects

Chang Su, Shou-Yang Zhai, Yu Cao, Xiao-Fei Deng and Chao-Jun Wu

Abstract This article analyses the difference between McWiLL system and paging system in indoor wireless coverage of the nuclear island. It argues the principles and conditions that should be satisfied when doing indoor wireless coverage in the nuclear island. Moreover, how to do design work of NI indoor wireless coverage under these principles and conditions is discussed. Finally, it finishes a proposal of NI indoor wireless coverage using McWiLL system and summarizes the main points of the proposal.

Keywords Nuclear island · Wireless · Indoor coverage

1 Introduction

It is difficult for wireless signal goes through because of many rooms, complex structure and thick wall in the Nuclear Plant. To achieve the purpose of the wireless indoor distribution basic universal coverage, the design is difficult and the program is complicated. Therefore, it's necessary to use electromagnetic theory and radio space propagation theory, combined with the currently identified the technical characteristics of the wireless system for specialized research, design solutions to meet the requirements of the NI of wireless indoor distribution.

C. Su (✉) · S.-Y. Zhai · Y. Cao · X.-F. Deng · C.-J. Wu
China Nuclear Power Design Co., Ltd., Shenzhen, China
e-mail: suchang@cgnpc.com.cn

S.-Y. Zhai
e-mail: zhaishouyang@cgnpc.com.cn

Y. Cao
e-mail: caoyu@cgnpc.com.cn

X.-F. Deng
e-mail: dengxiaofei@cgnpc.com.cn

C.-J. Wu
e-mail: wuchaojun@cgnpc.com.cn

Y. Xu (ed.), *Nuclear Power Plants: Innovative Technologies for Instrumentation and Control Systems*, Lecture Notes in Electrical Engineering 400,
DOI 10.1007/978-981-10-3361-2_13

2 Comparison with Reference Nuclear Power Plant

NI wireless indoor distribution of reference nuclear power plant Ling'ao Phase II Project adopts paging system (commonly known as beeper-beeper). The paging system works at the frequency band of 280 MHz. From the relation formula between wavelength and frequency: $\lambda = c/f$ (λ means wavelength, c means velocity of light, and f means frequency), we know that the wireless signal frequency is only 280 MHz. The signal frequency is low, wavelength is long, and diffraction easily occurs according to the physics and electromagnetic field theory. Therefore, indoor distribution by the paging system is still simple, even if NI indoor structure is complex. The coverage problem of the whole NI building will be solved, if several paging transmitters are put in the NI building. But the paging system is lag in technology, shrinking in market and difficult in equipment purchase, so it has been eliminated. New project will not use the paging system again, so the paging system can not be used for indoor distribution.

CPR1000 project NI indoor distribution plans to adopt McWiLL (Multi-Carrier Wireless Information Local Loop) broadband wireless access system. The system can completely replace the paging system in function. Nevertheless, the working frequency of the system reaches is 1800 MHz which is several times as high as that of the former paging system, the wavelength is short, diffraction does not easily occur, so the design difficulty of indoor distribution increases, in the face of daedal indoor environment of NI. McWiLL system is completely different from the paging system, so it can not design indoor distribution with reference to former paging system. It must start all over again, and restudy to obtain a feasible scheme which can meet the operating requirements.

3 Study on Indoor Distribution Scheme

3.1 Design Principles of Indoor Distribution

The selection of signal source for NI indoor distribution adopts connecting base station unit and single circuit in-line tower amplifier, and connecting ceiling antenna by means of connection with power divider or coupler. For the area with larger coverage area and far away from the base station unit, the network signal coverage adopts connecting optical fiber machine by remote at the back of base station unit.

As for the ceiling antenna, main control room and reactor center will not design the ceiling antenna, but other places in the room inside NI, namely all areas that people can touch will be covered with signal. The coverage of special areas such as long and narrow corridor, large-size electrical cabinet room, reactor periphery and reactor top-level maintenance area will adopt special antenna such as directional antenna. But most of rooms will install omnidirectional ceiling antenna to ensure seamless coverage of most of areas in NI.

3.2 Selection of Antenna Power

Table 1 Radiation Oscillation Disturbance Source and Immunity Scope [1] in Sect. 5.3 of GB/T 11684-2003 Electromagnetic Environment Conditions and Testing Procedures for Nuclear Instrumentation stipulate basic requirements of anti-disturbance requirements of nuclear instruments which are used in different places, as shown in the following table.

According to the requirements of Appendix B.1 of the Specifications, the immunity of nuclear instruments for nuclear power plant is 3. From the table, it can be seen that the immunity of wireless signal is 3 V/m (the frequency of McWiLL system is 1800 MHz).

The provisions on immunity requirements in other specifications are shown as follows:

- According to GB 17626.3-2006 Electromagnetic Compatibility—Testing and Measurement Techniques—Radiated, Radio-frequency, Electromagnetic Field Immunity Test [2], the equipment shall meet the test class 3 and test voltage 10 V/m.
- According to the provisions of Sect. 8.4.1.2 of GB 14048.1-2006 Low-voltage [3].

Switchgear and Controlgear—Part 1: General Rules, EMC test requirement for the product is 10 V/m.

Table 1 Radiation oscillation disturbance source and immunity scope unit V/m

Immunity	Disturbance source						
	9 kHz–27 MHz, any source	27 MHz frequency band, CB (citizens band)	Amateur radio, all frequency bands	27–1000 MHz portable, except CB	27–1000 MHz mobile, except CB	27–1000 MHz, except CB, portable, mobile	1000 MHz–40 GHz, all sources
A (controlled)	Consider item by item according to the equipment requirements						
1	0.3	0.3	0.3	0.3	0.3	0.3	0.3
2	1	1	1	1	1	1	1
3	3	3	3	3	3	3	3
4	10	10	10	10	10	10	10
5	30	30	30	30	30	30	30
X (harsh)	Consider item by item according to the situation						

- DL/T 478-2001 General Specifications for Static Protection [4], Security and Automatic Equipment stipulates that the ability of the device to withstand electromagnetic interference shall meet the provisions of GB/T 14598.9 Electrical Relays Part 22-3: Electrical Disturbance Test for Measuring Relays and Protection Equipment-Radiated Electromagnetic Field Disturbance Tests [5], namely 10 V/m.

To sum up, in consideration that other specialties fail to raise concrete number requirements for immunity of equipment in NI, from the angle of security, the equipment in NI shall at least meet the immunity requirements of not more than 3 V/m.

$$3\,\mathrm{V/m} = 20\lg 3\times 10^{6} = 129.54\,\mathrm{dBuV} \tag{1}$$

According to the reduction formula:

$$\mathrm{dBm} = \mathrm{dBuB} - 113 \tag{2}$$

$$129.54\,\mathrm{dBuV} = 129.54 - 113 = 16.54\,\mathrm{dBm}.$$

Because the antenna adopts the ceiling installation mode, it is impossibly close to the equipment of other systems in NI, so the power at 1 m away from antenna shall be firstly calculated.

According to indoor propagation model which is suggested by ITU-R Rec.P.1238 [6]:

$$Ls = 20\lg f + 10n\lg d + Le - 28 \tag{3}$$

(f means working frequency, unit MHz; d means path distance, unit m; n means indoor propagation index, a constant within 2.0–3.3, relating to the nature of the building; in consideration of the situation in NI, n = 3; Le means through-wall loss)

$$Le = 37 - 7.9\lg f = 11.28\,\mathrm{dBm} \tag{4}$$

Calculate the attenuation at 1 m away from the antenna:

$$Ls = 20\lg 1800 + 30\lg 1 + 11.28 - 28 = 48.4\,\mathrm{dBm}$$

The transmitting power at entrance of antenna is generally L(0) = 10 dBm, and we can calculate the power at 1 m away from antenna:

$$10 - 48.4 = -38.4\,\mathrm{dBm} < 16.54\,\mathrm{dBm}$$

which is far lower than the value required for resistance to electromagnetic radiation interference. Therefore, we take the transmitting power at entrance of antenna L(0) = 10 dBm, that is to say, 10 mW meets the requirements.

According to one-way communication characteristics of the system in NI, the mobile phone terminal will not transmit signal in NI, so we do not need to consider the interference and influence of signal transmission at mobile terminal on other important devices, but we shall guarantee the transmitting power at entrance of antenna less than 10 mW.

3.3 Confirmation of Signal Coverage Radius of Antenna

The signal coverage radius of antenna is set as r, the power of antenna at r is L(r), and loss is L(r).

In consideration that pipelines, portal frame, equipment cabinet and network structure equipment made of various metal materials in NI will shield and absorb wireless signal, there is possibly the disturbance of other wireless communication signals, so theoretically, there is excess loss of 20 dB, therefore:

$$L(r) = L(0) - Ls(r) - 20\,\mathrm{dBm} \tag{5}$$

According to technical features of McWiLL wireless system, mobile terminal under wireless environment of −90 dBm can meet basic communication requirements. Therefore:

$$L(r) \geq -90\,\mathrm{dBm}$$

$$L(0) - Ls(r) - 20 \geq -90\,\mathrm{dBm}$$

$$10 - (20\lg f + 10n\lg r + Le - 28) - 20 \geq -90$$

$$r \leq 11.3\,\mathrm{m}$$

For appropriate margin, preventing that insufficient signal power at the edge of antenna coverage radius will influence signal reception, we take r = 8–10 (m).

3.4 Field Test

For proving the reliability of data calculated theoretically, we conduct relevant tests in the constructed simulation test environment.

(1) Take the power at entrance of antenna as 10 mW;
(2) Test the strength of cell phone receiving signal within 3–10 m away from antenna by simulation test.

Test data is shown as follows (Table 2):

Table 2 Test data

Distance from test point to antenna (m)	The strength of cell phone receiving signal by simulation test (dBm)
3	−56.3 to −58.6
4	−58.4 to −60.2
5	−61.3 to −64.6
6	−62.8 to −65.6
7	−66.6 to −67.8
8	−69.3 to −72.6
9	−73.8 to −75.6
10	−76.8 to −78.2

Thus, it can be seen that, when power of antenna is 10 mW, the coverage radius of 8–10 m can meet basic communication requirements of cell phone terminal −90 dBm.

3.5 Arrangement of Indoor Distribution Antenna Point Location

According to the principles of the foregoing discussion, we design the arrangement of antenna point location, as shown in Fig. 1 (take NI 1LX −6.70 m layer as an example):

Among them, □ is the position of signal source, and is generally optical terminal equipment or in-line tower amplifier. Locating place is close to the position of terminal box of other communication subsystems. ⊟ is the punching position of cable, and the cable feeder connecting antenna shall traverse wall along the existing cable perforation of other communication subsystems to the greatest extent.

3.6 Laying of Feeder Cable and Average Power Distribution

After completion of antenna point location, it is required to lay feeder cable. The principle is to control roughly similar antenna transmitting power at each end, and meet the requirements of less than 10 dBm. This needs to adopt network topology which is composed of coupler or power divider to effectively distribute power. Power divider and coupler of different parameters complete equalization of antenna power under the premise of reasonable route so as to guarantee standard wireless signal edge field strength of coverage area [7].

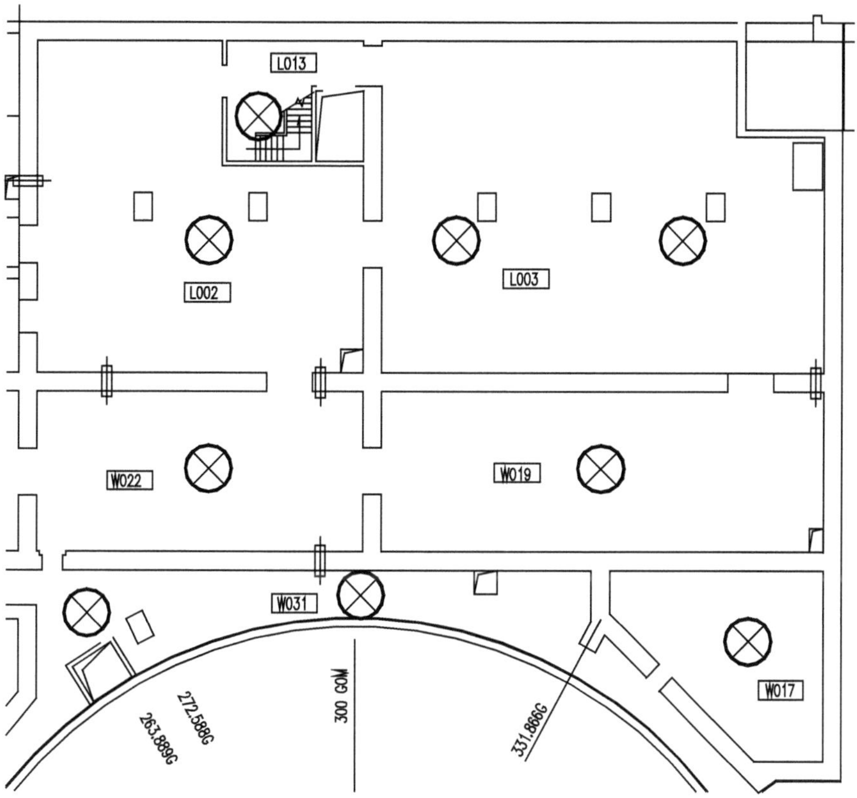

Fig. 1 Arrangement of antenna point location of NI 1LX −6.70 m layer

Table 3 Line loss of feeder every 10 m in each frequency band

Frequency	Feeder	Loss/10 m
1800 MHz	1/2 feeder	1.1
	7/8 feeder	0.7

The parameters of feeder, coupler and power divider are shown as follows (Tables 3 and 4).

The feeder loss is related to its length, so the change in loss arising from the change in feeder length must be considered during design of feeder cable route in order to ensure equalization of antenna power.

3.7 Design Results of Indoor Distribution

Indoor distribution design drawing is basically completed, after completing arrangement of antenna point location and laying of feeder cable as well as adding of equipment coding. The area in NI is too big, so we adopt the subarea design

Table 4 Devices adopted for indoor distribution and loss of each device

Name of device	Power loss value of coupling terminal	Power loss value of straight terminal (exit of power divider)
6 dB coupler	6	1.5
10 dB coupler	10	1
15 dB coupler	15	0.5
20 dB coupler	20	0.5
30 dB coupler	30	0.5
Two-port power divider	/	3.5
Three-port power divider	/	5.5
Four-port power divider	/	7.5

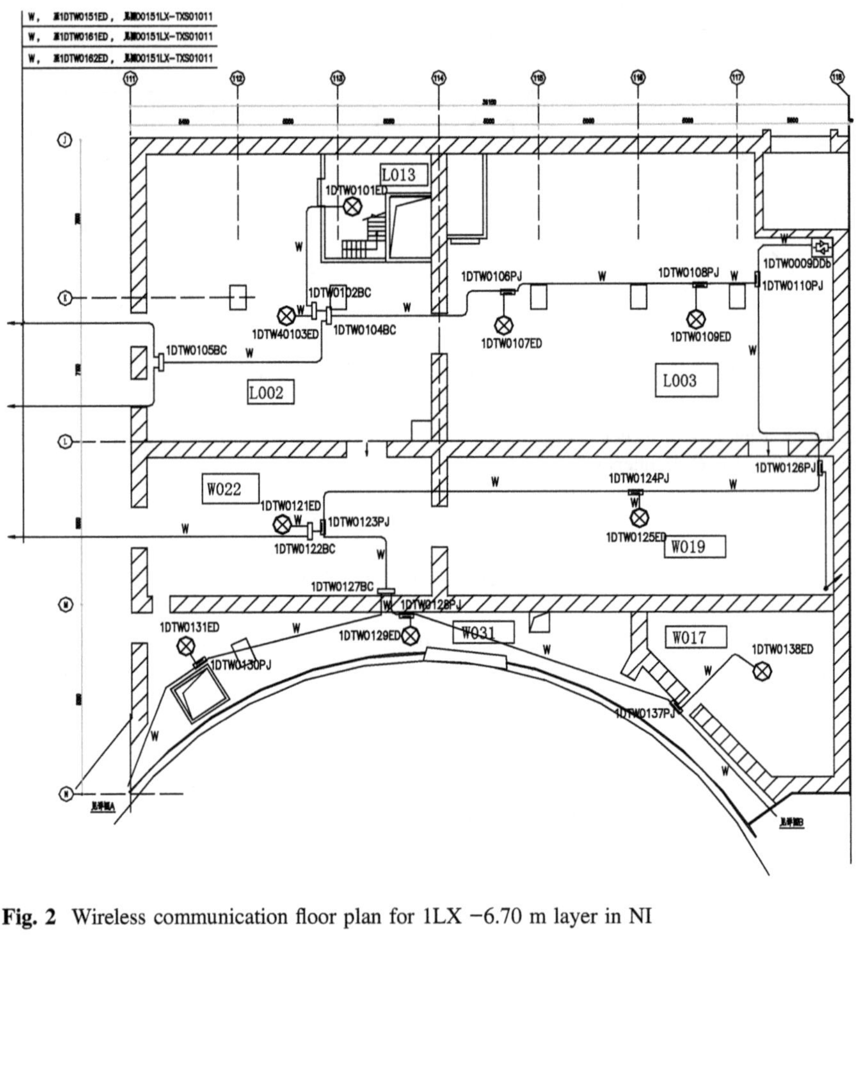

Fig. 2 Wireless communication floor plan for 1LX −6.70 m layer in NI

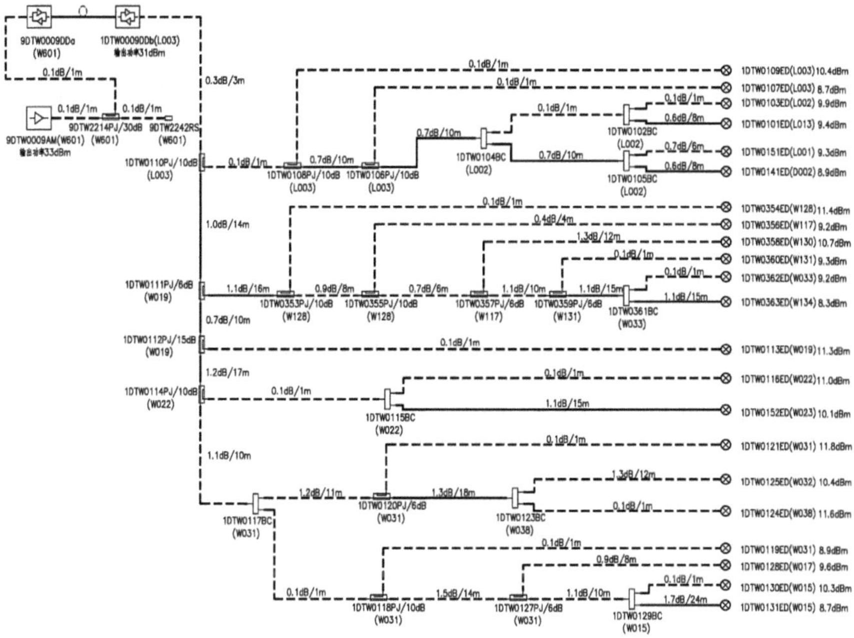

Fig. 3 Wireless communication system chart of NI

method during detailed design. The whole NI building is designed according to 1LX, 9LX, 2LX, 9NX, 1KX, 2KX, 1DX and 2DX subareas respectively. The drawings of different subareas are based on different layers. The following is the drawing for 1LX −6.70 m layer (Figs. 2 and 3).

3.8 Power Supply of Indoor Distribution System

For research and design of indoor distribution in NI, all devices are basically passive devices which do not need special power supply, including coupler, power divider, antenna and feeder. Active devices mainly include base station and optical terminal equipment, UPS power supply specially used for communication system in W601 communication equipment room provides the power to them so as to ensure the power supply reliability of the system, and avoid that the workers in NI fail to receive information timely in case of power failure arising from any accidents.

3.9 Summary of Indoor Distribution Design

- The power at entrance of indoor antenna is controlled at 10 dBm. The antenna or cable may shift in a small range during installation, so we consider a certain margin during design.
- The coverage radius of omnidirectional ceiling antenna is considered as 8–10 m.
- Elevator shaft and staircase adopt directional antenna.
- The adjoining room with the wall thickness of 200 mm or less is equipped with 1 antenna according to the room size and position. Each room with the wall thickness of more than 200 mm is equipped with antenna.
- For avoiding the influence of antenna on the instrument control equipment and other important equipment, the main instrument control equipment room, control room and storage battery room are not equipped with antenna.
- All base stations in NI are set in W601 communication equipment room of +15.5 m layer, appropriate clearance around base station shall be reserved for the convenience of heat dissipation of base station.
- In principle, the optical terminal equipment at each layer shall be installed near terminal box of other subsystems.
- In principle, all cables shall traverse former communication holes (horizontal, vertical). The cables shall not be laid with wind pipe and other metal pipes in parallel. The radius of cable bending (if required) shall meet the requirements of cable technology specifications. Add galvanized steel pipe for cable laying when traversing the hole.
- Coaxial cable entering each diesel factory building shall enter from one place. The diesel factory building does not consider active devices.

4 Conclusions

This research proposes a solution for wireless indoor distribution in NI by adopting McWiLL broadband wireless access system. The scheme avoids the influence on important sensitive devices such as DCS inside NI, effectively solves wireless communication problem inside NI, and also opens new thought for wireless indoor distribution in NI after the paging system is eliminated.

At the present stage, no organizations at home and abroad conduct design and research on wireless communication coverage in NI in combination with McWiLL system and nuclear power plant, which has important innovation significance. The research scheme can be applied to CPR1000 new projects. At present, it has been implemented in Hongyanhe, Ningde and Yang jiang nuclear power plants. Because of flexibility and universality, the appropriately improved scheme is also applicable to nuclear power plant of AP1000 reactor type, enjoying broad development prospect.

References

1. GB/T 11684-2003 Electromagnetic Environment Conditions and Testing Procedures for Nuclear Instrumentation.
2. GB 17626.3-2006 Electromagnetic Compatibility—Testing and Measurement Techniques—Radiated, Radio-frequency, Electromagnetic Field Immunity Test.
3. GB 14048.1-2006 Low-voltage Switchgear and Controlgear—Part 1: General Rules.
4. DL/T 478-2001 General Specifications for Static Protection, Security and Automatic Equipment.
5. GB/T 14598.9 Electrical Relays Part 22-3: Electrical Disturbance Test for Measuring Relays and Protection Equipment-Radiated Electromagnetic Field Disturbance Tests.
6. McWiLL Broadband Wireless Access Technology and Application, Wen Bin-2009-the People's Posts and Telecommunications Press.
7. Evolution of McWiLL: Industry Informatization-Oriented Broadband Multimedia Cluster System, Jiang Hui, Xu Ruifeng, Zhang Dandan, Hu Zhenxing-Telecommunications Science, 2008, 24(8).

References

Evaluation System of Software Concept V&V About the Safety Digital I&C System in Nuclear Power Plant

Peng-Fei Gu, Wang Xi, Wei-Hua Chen and Su-Yuan Yu

Abstract Since the digital technology was used in the safety digital I&C (Instrument and Control) system of nuclear power plant (NPP), its safety and reliability have been one of the most important factors to the safety operation of NPP. V&V activity is a significant method to insure the safety and reliability of the nuclear power I&C system software, the system to evaluate the efficient of V&V activity need further research. This paper on the basis of CPR1000 NPP Reactor Protection System (RPS) software development project, use the concept V&V activity for example, make a discussion for the evaluation of V&V activity, including the description of general tasks, definition of the Failure Modes, and analysis of V&V activity by FMEA (Failure Modes Effects Analysis). The results of our evaluation method in project show that the efficiency of V&V activities has been improved, and provide references for the evaluation to other NPP I&C system software development V&V activity.

Keywords Nuclear power plant · I&C · V&V · FMEA

1 Introduction

As the active development strategies for nuclear power, the installed capacity of which will reach 70–100 GWe at the time of 2020, account for more than 4% of the total installed capacity in China, and the proportion should be at least added to 6%

P.-F. Gu (✉) · S.-Y. Yu
Institute of Nuclear and New Energy Technology, Collaborative Innovation Center of Advanced Nuclear Energy Technology, Tsinghua University, Beijing, China
e-mail: gupengfei@cgnpc.com.cn

P.-F. Gu · W. Xi · W.-H. Chen
Instrumentation and Control Department, China Nuclear Power Design Co., Ltd., Shenzhen, China
e-mail: wang.xi2@cgnpc.com.cn

W.-H. Chen
e-mail: chenweihua@cgnpc.com.cn

Y. Xu (ed.), *Nuclear Power Plants: Innovative Technologies for Instrumentation and Control Systems*, Lecture Notes in Electrical Engineering 400,
DOI 10.1007/978-981-10-3361-2_14

according to the Low Carbon (LC) plan with the installed capacity arriving at 150–200 GWe in the year of 2030. The time limit for a complement construction of NPP is more than 5 years, which means there are 10 nuclear power units for each year from now to the year of 2025. A mature, quantity production stage for the development of nuclear power is coming for us.

The sustainable development of nuclear power station not only relies on the acceptance of the nuclear economy in public, the improvement of the safety and reliability in nuclear technology is also more significant. The nuclear Digital Control System (DCS) is one of the most important devices for the safety of nuclear power station. Software is the kernel of DCS, which on the basis of CPU, to achieve the protection and logic of the devices for nuclear power station. The safety of the software affects the safety, reliability and economy of the NPP directly. Therefore, a strict verification and validation (V&V) activity for the whole lifecycle of software development is necessary [1, 2], and the method to evaluate the effectiveness of the V&V activity should be further discussion.

This paper arranged as follow. Section 2 introduces the work of V&V activity for DCS software development in NPP, and propose the general tasks and key points for the evaluate of concept V&V; Sect. 3 makes definitions for failure modes including the general tasks and key points; Sect. 4 combines to the CPR1000 project, bases on the constructed failure modes, use FMEA method to analyze and improve the V&V activity; Sect. 5 make a conclusion for the whole paper.

2 V&V Activities and Tasks

V&V activity is an important method to assure the quality of software. Verification conforms to requirements for all activities during each life cycle process, satisfies the standards, practices, and conventions during lifecycle processes. Validation satisfies system requirements allocated to the products at the end lifecycle activity, solves the right problem, and satisfies intended use and user needs in the operational environment [3]. V&V activity aims to locate and recognize the default or errors in the software, assure the correct process of software development, make the products satisfy the whole requirements from user, and assure the consistency of computer software and the technology requirements, make sure of the software functions correctly in the environment designed previously [4].

2.1 V&V General Tasks

According to HAD 102/16 (2004) [5], IEEE1012-2004 and IEC60880 [6], the V&V activities for DCS in NPP process by 6 stages, including management process, acquisition process, supply process, development process, operation process, and maintenance process. The most significant stage among them is the development process, in which the main V&V activities including Concept V&V, Requirements V&V, Design V&V, Implementation V&V, Integration test V&V, Installation V&V and checkout V&V. For each V&V activity, there are tasks to be complemented, named general tasks.

This paper use concept V&V as example to introduce the general tasks.

2.1.1 Concept Documentation Evaluation

Concept documentation evaluation insures that the concept documentation satisfies use requirements and complies with the precede needs, assures the restrains of interfacing systems and the imposed restrictions on provided approach, make analysis on system requirement and ensure the needs from user, including system function, end-to-end system performance, operation and maintenance requirements and so on.

2.1.2 Traceability Analysis

The traceability analysis implemented as follow. Firstly, make identification for the whole system needs, which should be accomplished completely or partially by software. Then, verify that precede needs can be traced by the system requirements. Finally, the traceability analysis starts between the software requirements and system requirements.

2.1.3 Requirements Allocation Analysis on Hardware, Software and User

The analysis verifies the completeness, correctness and accuracy of the concept requirement that has been allocated to hardware, software, and user interfaces for user needs. The completeness verifies that user needs should be satisfied by follows, including failure detection, isolation, diagnostic, and error recovery. The correctness verifies hardware, software, and user interfaces have been allocated to those performance requirements that satisfy the needs from user. The accuracy including the verification of the specification of external and internal interfaces for interface protocols, data formats, and the frequency of data exchange.

2.1.4 Hazard Analysis

Hazard analysis analyzes the potential dangerous to and from the concept system, including the identification of the potential system hazards and mitigation strategies for each hazard, the accession of the severity and the probability of each hazard.

2.1.5 Risk Analysis

The risk analysis including two parts, the identification of the technical and management risk, the proposed suggestions to mitigate or decrease the risks.

2.1.6 Security Analysis

The security analysis including: review the acceptable level of security, and then ensure confidentiality, integrity, availability, and accountability. Surely the risk related to system interfaces should be analyzed.

2.1.7 Criticality Analysis

The main activities of criticality analysis pay attention to the integrity levels [7], make sure that software integrity levels have been established for detailed functions, software modules, requirements, subsystem, or other software partitions; make verification of the assigned software integrity levels to be correct; insure the software integrity levels have been assigned to individual software components. The assignment of the software system should be the same as highest level assigned to any individual element; the assignment of software component should be the same or higher than the software integrity level, while any software component that can influence individual software components are assigned a higher software integrity level.

2.2 V&V Key Points

The concept V&V activities should pay more attention to the "key points", which reflect a specific consideration for the nuclear power station engineering project. An evaluation of whether these "key points" has been considered in software V&V activity could show further effectiveness of V&V works. This paper discusses the importance of DCS Contracts and the Requirement Tracing on the basis of the V&V activity in CPR1000 RPS DCS software development project.

2.2.1 DCS Contracts

The DCS contracts should be included in the input documents as the reference files that can be traced.

Theoretically, the RPS specification is the baseline of concept V&V activities. With the progressively implementation of the construction of NPP, the technical details in DCS contracts, which signed at the beginning of engineering project for the consideration of the whole schedule, may not in consistent with the RPS specification, these inconsistent points cannot satisfy the original requirements of the NPP. Therefore the analysis of technical points is necessary.

2.2.2 Requirement Tracing

As the part of the requirement management [8], the requirement tracing establish traceability links [9] between every neighboring stages to provide a foundation for requirement modification management and version control. As the NPP DCS RPS software lifecycle has covered each stage including development, operation, and maintenance and so on, the mistakes and errors discovered in V&V activity not only beneficial to the quality assurance, and also provide a convenient for the Experience feedback of succession operation and maintenance if the NPP need renovation. Therefore, the requirement tracing matrix should be established exactly at the beginning of concept V&V.

3 Definitions of Failure Modes

Failure modes means the system or its sub-system or components do not satisfy their design or function of system requirements, as the example of the concept V&V illustrated in Table 1, Failure modes can be classified into five parts, including the inconsistent of legislation and standards, Lack of general tasks, Unaccomplished plan, inconsideration of the specific in engineering project, and Uncompleted requirement tracing. The definitions can be used for the classification of the Failures events that discovered by V&V activity.

Table 1 Definitions of failure modes in concept V&V activity

V&V activities	Failure modes	Definitions
Concept V&V	Inconsistent of legislation and standards	The V&V tasks are not in consistent of legislations and standards such as HAD102, IEEE1012
	Lack of general tasks	Lack of general tasks, such as hazards analysis
	Unaccomplished plan	The formulated plans for V&V activities haven't been totally accomplished
	Inconsideration of the specific in engineering project	The DCS contracts are not included in input documents
	Uncompleted requirement tracing	The requirements are not arranged to entries and tracked by using a requirement tracing tool

4 Failure Modes and Effect Analysis

The project, of which this paper on the basis, uses FMEA (Failure Mode and Effect Analysis) method to do an effectiveness analysis for concept V&V activities. In this project, according to the FMEA result for the first round concept V&V activity and the project schedule, the V&V teams formulate improvement measures for the second round concept V&V activity, and according to the FMEA result for the second round concept V&V activity and the project schedule, V&V teams formulate improvement measures for the third round concept V&V activity.

The FMEA results of the concept V&V activity in the project are described in Table 2, the "Failure results" means an uncompleted tasks results in V&V for the reason of the corresponding "Failure Modes"; The "Effect Analysis" evaluates the influence on the effectiveness of V&V activities by analyzing the failure events; "Improvements" assure the effectiveness of V&V activities by formulating improved method; "Problem Classifications" illustrated in Table 3, which defines the different levels of problems according to influences on V&V activity.

As showed in Table 3, to solve the failure modes, the backward tracing is done in the second round concept V&V activity; The requirement tracing tool named DOORS [10] is used in the second round concept V&V activity to establish the requirement tracing matrix; The FMEA method is used in the second round concept V&V activity and product the reports for the hazard analysis; The DCS contracts are included in the foundation documents for analysis in the third round concept V&V activity. Compared to the first round, the second round concept V&V activity discovered other 51 mistakes or errors, and the third round concept V&V activity find other 34 questions, by use the FMECA (Failure Mode Effect and Criticality Analysis) [11] method, an enhanced effectiveness of V&V activity has been showed in project results.

Table 2 Problem classifications

Problem levels	Discretions
Serious	Events will lead to the failure of the whole V&V activities.
Important	Events will result in losing of important contents of V&V activities
General	Partly influence on V&V activities and need pay further attention to the analysis of effectiveness for V&V activities
Suggestion	No influence on the effectiveness of V&V activities, but may lead to unfavorable operations and maintenances in the future

Table 3 FMEA results of concept V&V activities

No.	Failure modes	Failure results	Effect analysis	Improvements	Problem classifications
1	Lack of general tasks	The backward tracing is not done in the first round concept V&V activity	Ignoring of the extra requirements in DCS design file, and result in the failure of V&V activities that extra functions have been designed in system	The backward tracing is done in the second round concept V&V activity	Serious
2	Uncompleted requirement tracing	The requirement tracing is not done in the first round concept V&V activity	It is difficult to implement the requirement tracing in the succession V&V activities	The requirement tracing tool named DOORS is used in the second round concept V&V activity to establish the requirement tracing matrix	Suggestion
3	Unaccomplished plan	The hazards analysis, which is formulated in the plan, is not done in the first round concept V&V activity	The mistakes and errors cannot be found in system structures and the safety of functions cannot be analyzed	The FMECA method is used in the second round concept V&V activity	Important
4	Inconsideration of the specific in engineering project	The DCS contracts are not included and analyzed in the foundation documents in the second round concept V&V activity	The technical details in the DCS contracts that inconsistent with the RPS specification may not be found and result in lack of analysis for the differences	The DCS contracts are included in the foundation documents for analysis in the third round concept V&V activity	Important

5 Conclusions

On the basis of CPR1000 NPP RPS software development project, this paper gives a further discussion for V&V evaluation system, use FMEA system method to analyze the failure modes and results in V&V activities and provide improvements. By using the improvements in succession V&V activities, an enhanced efficiency of V&V activity has been showed in results of project, provides an reference for the evaluation of other NPP I&C software V&V activities.

Acknowledgements This project was financially supported with funds which provided by a reliability research of reactor protection system which based on digital microprocessor and electrical equipment (national science and technology major project). The award number is 2014ZX06004002-004.

References

1. Liu Z, Jiang G J, Sun Y B (2011) The V&V activities and techniques for safety-class I&C system in the nuclear power plant [J]. Chinese Journal of Nuclear Science and Engineering, 31(2):45–50.
2. Software Engineering Standards Committee of the IEEE Computer Society (2004) IEEE 1012 IEEE Standard for Software Verification and Validation [S]. Institute of Electrical and Electronics Engineer, New York.
3. R.G. 1.152 (2006) CRITERIA FOR USE OF COMPUTERS IN SAFETY SYSTEMS OF NUCLEAR POWER PLANTS [S]. U.S NUCLEAR REGULATORY COMMISSION.
4. International Electro technical Commission (2006) IEC 60880 Nuclear power plants-Instrumentation and control systems important to safety-Software aspects for computer-based systems performing category a functions [S]. International Electro technical Commission, Switzerland.
5. HAD 102/16 (2004) Safety of Nuclear Power Plant Design Regulations Guides [S]. Doctoral dissertation.
6. Nuclear Power Engineering Committee of the IEEE Power Engineering Committee (2010) IEEE 7-4.3.2 IEEE Standard Criteria for Digital Computers in Safety Systems of Nuclear Power Generating Stations [S]. Institute of Electrical and Electronics Engineers, New York.
7. Ding Y X, Gu P F, et al. (2015) Study on Standard about Safety Digital I&C System in NPP [J]. PROCESS AUTOMATION INSTRUMENTATION, 36(11):61–64.
8. Xia D Y, LIU W P (2015) Requirement Management on Nuclear Power Plant DCS for Development [J]. INSTRUMENTATION, 22(2):63–66.
9. Xiao W (2012) An Optimized Method for Software Requirement Development Process Using Doors [J]. Computer Application and Software, 29(9):175–177.
10. IBM Corporation (2010) Rational DOORS Help [CP/DK], IBM Corporation.
11. Wu X K, Wang G S (2014) Nuclear Safety Digital I&C System Software V&V Technical Standards Research [J]. Nuclear Standard Measurement and Quality, (4):16–22.

Discussion About Issues of Human Error in Digital Control Room of NPP

Yan Feng, Zhong-Qiu Wang, Jing-Bin Liu and Yin-Hui Guo

Abstract This article defines and classifies human errors simply, and also analyzes representative human errors in nuclear power plant. This article also describes the difference between digital control room and traditional analog control room of nuclear plant, such as composition, information display, procedure, alarm system and operate mode. This article introduces human characteristics in digital control room including the role of the operator, the task load of operators, operator's ability and experience, and digital human system interface. A example is given for human issues in digital control room. And the article simply studies and discusses preventive countermeasures in digital control system of nuclear plant at last. Since in preventing the human errors have common measures, but are also very different, much research work has been carried out.

Keywords Digital control system · Nuclear power plant · Human factor engineering · Human error

1 Introduction

Since nuclear power plant is a complex man-machine system, its design, operation and management must depend on people. With the development of science and technology, the performance of the hardware equipment of nuclear power plant is

---

Y. Feng · Z.-Q. Wang · J.-B. Liu (✉) · Y.-H. Guo
I&C Department, Nuclear and Radiation Safety Center, Beijing, China
e-mail: liujingbinjob@163.com

Y. Feng
e-mail: fengyan@chinansc.cn

Z.-Q. Wang
e-mail: wangzhongqiu@chinansc.cn

Y.-H. Guo
e-mail: Gyh86126@126.com

Y. Xu (ed.), *Nuclear Power Plants: Innovative Technologies for Instrumentation and Control Systems*, Lecture Notes in Electrical Engineering 400,
DOI 10.1007/978-981-10-3361-2_15

increased, also the operation environment and the reliability of nuclear power plant is greatly improved. Although the nuclear power plant is improved by the degree of automation, then its design, operation and management as well as the specific operation also depend on people, therefore people plays more and more important. Because that people's psychology, physiology and education are different, it is possible to arise human errors and to affect the safety of nuclear power plant. Instrumentation and control system is changed from the traditional analog control to the widely used digital control technology, then the man-machine interface, procedures and cognitive behavior model of operator change, so new error mechanism occurs [1]. Therefore, it is important to carry out the digital control system of human error research, to find out the underlying causes, and then to put forward prevention strategies.

1.1 Definition and Classification of Human Errors

Human error is that the result of people's behavior deviates from the specified target, or exceeds the acceptable limits, and has a bad effect. Different experts and scholars define human error from different angles. Swain think that human errors are people's behavior or action exceed the scope in any more than a certain standard system to accept the normal work of the accepted standard [2]. Reason defined human error from the point of view of psychology. He thinks that errors are all such phenomenon that people despite a series of planned psychological or physical activity, but did not achieve the expected results. This failure cannot be attributed to certain external factors involved in [3]. Lorenzo think that if role in system of any behavior (contains no enforcement or lax enforcement of its behavior) beyond the system of tolerance, it is human error [4].

But regardless of how they are defined, human errors have the following characteristics: repeatability of human errors; the potential and irreversibility of human errors; error often derived by situational context; the inherent variability of people's behavior; reparability of human errors [5].

The International Atomic Energy Agency divides human errors into three major categories [6]:

Class A: the human errors can result potential failure of a device or system.

Class B: the human errors or the combination of equipment failure directly led to the occurrence of the beginning events.

Class C: the human errors cause after the fault occurs, and they may be errors in the implementation of safety behavior or the action to deteriorate the consequences.

In addition to the International Atomic Energy Agency's human error classification method, there are three other kinds of classification methods: namely, the behavioral method, the relational method and the conceptual method. The behavioral method is only related to the behavior of the observed and unexpected and focuses on what behavior occurs. The relational method emphasizes the cognitive mechanism of human error, and tries to provide a framework for improving system design and training program. The conceptual method is one of the most effective and most useful method, which divides all human error into slip, lapse, mistake [7].

1.2 Difference Between the Digital Control Room and the Traditional Control Room

The traditional control room is mainly composed of MIMIC panel, alarm window, control instrument, recorder, control panel, control button and so on. The control console is in the middle of control room, the equipment group is distinguished by physical distance, the contour line or the color code, the control switch is distinguished by using the different size and the shape. Overall, the overall layout is more intuitive. The operator can easily find all kinds of information of main control room and determine the severity of alarm according to the color and position of alarm.

The layout of digital control room is relatively simple, and there are four main computer operator workstations, large screen display and backup panel. Each workstation includes visual display unit, mouse and the track ball. The information of the whole nuclear power plant is obtained from the computer, and the information is much larger than that of the traditional control room, but it is not intuitive in the traditional control room.

Digital control room can provide more comprehensive and accurate nuclear power plant information, and this helps the operator a lot. For example, the operator can quickly find fault causes through the alarm. Digital control room can automatically judge in the handling of the accident, and then the burden of operator is reduced. Computerized procedures system, advanced alarm system and graphical information display system play a positive role for the team's performance, reduce the work load and reduced human errors [8].

Although, the digital control room also increases the work load of the operator. Because of the huge amount of information, the operator obtaining the desired information from the computer is not as easy as the traditional main control room. To obtain the required information, operator must be familiar with the man-machine interface and need to open different screens and windows. It is a challenge for the operator to obtain the key information in a short time since a large amount of information exists. The computerized procedures system also changed the traditional mode of operation of operators. This may also induce new human errors, such as pattern confusion, data entry errors, loss of situational awareness, etc.

Table 1 describes the difference between the digital control room and the traditional control room.

In summary, the digital control room changes in the system, man-machine interface, procedure system, alarm system, analysis and decision support system, the team structure and communication path. The cognitive processes and behaviorial mode of the operator are also changed. The man-machine interaction is more

Table 1 The difference between the digital control room and the traditional control room

	Digital control room	Traditional control room
Composition	Computer operator workstations, large screen display and backup panel	MIMIC panel, alarm window, control instrument, recorder, control panel, control button
Information display	The amount of information is huge	The information is intuitive
Procedure	Computerized procedure	Printed procedure
Alarm system	To spend more time in query the alarm	The display of alarm system is very intuitive
Operate mode	Software-based control	Hardware-based control

closely and more complex in the digital control room. The role of operator is from the traditional patrol plate, monitoring and operation gradually changed to monitoring and decision, which makes human errors appear new characteristics and new human error distribution, human errors data and failure mechanism.

2 The Human Characteristics of the Digital Control Room

The figure below is the overall DCS structure diagram. Automation level, communication level, process information and control level are the focus of attention in safety review.

The operator has to cope with the very complex system in the control room of nuclear power plant. In the traditional analog control room, the operator needs to move back and forth, to read the information from the larger control panel and to operate the button. In the advanced digital control room, the reliability of system and equipment is improved. The operator can monitor and control nuclear power plant by using the integrated system. The human characteristics of the digital control system mainly are the following aspects:

(1) The role of the operator. The operator is mainly monitoring and operating system in traditional control room, and the operator may mostly monitor and make decision in digital control room. The operator's task contains more cognitive work and the operator performs cognitive tasks through a series of cognitive behavior.
(2) The task load of operator. The information through the traditional simulation system displayed is very intuitive. However, the digital control room is based on the large screen display and the computer workstation display information, and displays the huge amount information. If the operator needs to obtain information, he has to configure screen, navigate, and check, also need continuous navigation. These have increased the task load of operators.

(3) Operator's ability and experience. Digital control system is a complex system, the higher the degree of automation, the more complex the system. This require the operator's ability and experience.
(4) Digital human system interface. The digital human system interface mainly includes three aspects: computerized procedure systems, information display and control based on VDU, alarm system.

① Computerized procedure systems. The printed procedure is used in the traditional control room, while the digital system is based on computerized procedures. Also computerized procedure systems is the complex structure.
② Information display and control based on VDU. The indicator in traditional control panel is usually cured, however, the display based on computer information is not limited to physical space and can be displayed through the screen of arbitrary configuration information. The same information in different screen position may not is the same. Therefore, the difficulty of operator positioning, search and identification is increased. In addition, soft control is more complicated than traditional control in the control room, soft control may sometimes induce mis-operation.
③ Alarm system. The display in traditional alarm is very intuitive. According to the color and location, experienced operators can judge the severity of the alarm. The operator may need to query, filter and other acts to query the alarm in the computer alarm. In other words, the operator needs to spend more time to search and confirm the alarm, and then may delay other tasks.

A example about issues of human factor in digital control room is as followed.

IRS 8023 (NRC Information Notices 2009–09 and IN 2008–13) covers events relating to improper flow controller settings in the injection systems at several BWRs, which resulted in system-flow oscillations rendering the systems inoperable. Testing failed to identify this because the systems' alignment during surveillance differed from that when they are called upon to perform their safety functions.

An unplanned automatic reactor trip occurred at a BWR in response to a turbine control valve fast-closure signal caused by failure of the digital feedwater control system (DFWCS). The root cause was a defect on the 24 V direct current converter board of both the primary and secondary power supplies. It was established that the primary power supply was in degraded condition and the secondary power supply had failed completely. This resulted in erratic performance of the input and output boards. Also, the licensee had previously received two precursor alarms pointing to possible problems with the digital power supplies, but no troubleshooting had been carried out.

Licensees should ensure that main feedwater system digital modifications are fully understood and properly implemented and that operators are trained in the modified system and abnormal operating procedures in the event of feedwater system failures.

Operating experience highlights the importance of timely investigation, troubleshooting and analysis of power supply and communications alarms. Operators should be trained in the monitoring of important parameters such as power supply voltage to ensure the early detection and correction of problems.

3 The Response Measures to Human Error in Digital Control System

The causes of human errors are divided into internal and external factors. The external factors refer to the individual work environment, and the internal factors include the physiological characteristics and psychological characteristics of the individual. Weitzen summarize three causes of human errors, that is overload, mistaken decision making and man-machine factor [7].

Overload is that a person's ability in a certain mental state does not adapt the physical, physiological and psychological load. the person's ability refers to affordability of the physical, physiological and psychological aspects (human nature); current mental state; and current knowledge and the technical level related work; the temporary ability decline due to taking drugs or alcohol, stress, fatigue. Mistaken decision is that the choice of unsafe behavior is more logical than the choice of safety behavior in some times. Ergonomic reasons mainly include two aspects: the current working conditions don't adapt to person's physique, and the work platform design makes people easy to make mistakes.

A lot of statistical practices show that a lot of major security incidents and accidents of nuclear power plant are caused by human errors. Therefore, it is particularly important to strengthen the nuclear safety culture and to prevent people from mistakes. The main measures for reducing human error are the design of advanced control room, the use of human factors engineering principles, in addition to strengthening the training of education, the management and the working environment in view of the digital control system.

1. The nuclear safety culture is in the field of nuclear industry culture, and strengthening nuclear safety culture and preventing human errors can not depend on alone a person or unit. It refers to government and operating units.
   (1) The government: the government and its departments should formulate relevant laws and regulations, strengthen law enforcement on the basis of existing laws. Regulatory authorities have complete safety regulations, guidelines and related documents, supervising nuclear facilities independently and evaluating nuclear facilities regularly.
   (2) Operating units: operational units have main responsibility to deal with nuclear security and to prevent of human errors. Operating units should pay attention to safety, adjust the relation between the power management organization and government supervision departments, review operating

units regularly and strengthen the training of personnel in power plant. Operating units also should organize the on-site work and strengthen the education of safety culture. Nuclear safety education includes nuclear safety laws and regulations, nuclear safety knowledge, nuclear safety skills education and nuclear safety attitude training. The operation personnel should consciously abides by the regulations, develop a rigorous style of work, and improve the capacity of judgment, prediction and treatment only through the nuclear safety education and training.

2. Strengthening safety management and forming human factors analysis and experience feedback system.

The possibility of human error can be reduced by strengthening the safety management. There are many measures for strengthening safety management, such as implementing work permit system, ensuring that staff qualification and technical level comply with operating conditions, and ensuring the operation in a fully prepared and adequate safety measures. At the same time, it is necessary to improve the working conditions and the environment, to establish appropriate safety monitoring measures, to monitor the high risk operation that it may lead to major accidents, to correct timely the mistakes and to avoid causing serious consequences.

It is very important for analysing human events and forming experience feedback. The general practice is using the scientific safety management method and technology, changing passive "accident treatment" to the active "accident prediction", taking the active defense strategy, and reducing or avoiding the occurrence of human error as far as possible.

3. Defense in depth

Through the establishment of multi-layer overlapping security system to constitute a multi-line of defense, even if the failure of a certain line of defense can also be made up or corrected by other defense. To construct of active human accident defense in depth system, it is necessary to take a combination of technical, organizational and cultural measures to reduce and prevent from the management decision-making, organization, technology, accident analysis and reduction and feedback. The initiative to exploration and recognition of potential accident and to take measures to reduce an prevent accident are very important.

4. Improving the man-machine interface

Human errors may be caused because that the man-machine interface design is unreasonable, and adequate security and protection facilities is lack. It should establish the necessary physical barriers under the guidance of cost-effectiveness of the budget and effective prevent effectively human errors. The following aspects In the man-machine interface design should be paid attention to: (1) Reasonable matching man, machine and environment system, allowing the operator to easily and accurately obtain information, and correcting implementation of the related operations; (2) Using design preventing human errors, to ensure that operator operates in the right way; (3) It is necessary to provide sound, light alarm signal; (4) It is necessary to consider the necessary safety devices.

5. Application of human factors engineering principles

Human factors engineering study how to use the human's features systematically (including psychological and physiological characteristics) during the product or system development process. Using human factors engineering principles in nuclear power plant design can reduce human errors of endangering the safety and operability and improve plant safety.

In order to minimize human error, it should take full account of the human factor, retaining positive features and avoiding weaknesses of person in designing digital control room. For example, the functional analysis and assignment should consider handling capability and capacity between operator and the instrumentation and control systems. The allocation of functions to personnel and machine resources can take advantage of human and machine strengths and avoid human and machine limitations. Screen design should be full use of human factors engineering principles, such as font size, color, and other information packets must take into account the human factor. Alarm design should be graded according to the degree of importance. Alarm design needs to consider the design of alarm classification, filtering, suppression, so that the operator can quickly obtain the alarm he needed. Workstation design should provide more comfortable operation in sitting or standing, such as operator workstations disc table size, the size of the operator's visual angle. Digital control room has to provide a safe and comfortable operating environment, which can reduce operator fatigue and stress, and avoid mistakes. For example, it is necessary to consider the design of the main control room ventilation, noise design and lighting design.

6. Add human factor topics review

Human factor topics review may early detect issues of human factor, and take measures to reduce human errors.

4 Conclusion and Suggestion

Human error is one of the main factors causing accident. Especially after the Three Mile Island and Chernobyl nuclear accidents, people have a more profound understanding of this. Complex man-machine interface and computerized procedures are adopted in digital control room, which are very different from traditional control room. Both in preventing the human errors have common measures, but are also very different. This article only studied human error related content in digital control system of nuclear power plant, and we still have to carry out research work, to discuss strategies to prevent human error, and to promote the development of nuclear power.

References

1. ZHANG Li, WANG Yi-qun, HUANG Shu-dong. System Model for Preventing in Depth the Accident due to Human Factor (J). China Safety Science Journal. 2002, 12(01).
2. Swain A D, Guttmann H E. Handbook of human - reliability analysis with emphasis on nuclear power plant applications, 1983, NUREG/ CR 1278.
3. Reason J. Human Error UK. Cambridge: Cambridge University Press, 1990.
4. LI Peng-cheng. Human Error Cause Analysis and Application Study [M]. University of South China, 2006.
5. LIAN Shi-qian, ZHANG Li, WANG Yi-qun, SONG Hong-tao. Study on Human Error and Causal Factors (J). Industrial Safety and Environmental Protection. 2007, 33(11).
6. International Atomic Energy Agency. Human Reliability Analysis in Probabilistic Safety Assessment for Nuclear Power Plants [R]. Vienna: IAEA, 1995.
7. XIAO Guo-qing, CHEN Bao-zhi. Study on the Mechanism of Human Error and Its Reliability (J). China Safety Science Journal. 2001, 11(01).
8. LI Peng-cheng. Study on Human Error and Reliability in Digital Control System of Nuclear Power Plant. University of South China, 2006.

References

Improving Human-System Interface Design Through Human Behavior Assessment in the Control Room of Nuclear Power Plants

Xiaolu Dong, Dan Pan, Zhizhong Li, Yunbo Zhang, Yan Feng and Jingbin Liu

Abstract Human plays an important role in sophisticated and safety-critical systems such as Nuclear Power Plants (NPPs). Digital systems are applied in the control room alongside the development of technology. Operators in the control room of NPPs can interact with the system through traditional consoles as well as computer workstations. In order to evaluate the design of human-system interface, measures can be taken by assessing the task results and human performance. By observing the behaviors of operators, evaluators can identify potential problems of Human-System Interface (HSI) and thus making improvement. This paper briefly summarizes issues of human activities and performance assessment in NPPs, and then discusses the relationship between human behavior and HSI design. It provides some recommendations on interface design based on the observation of human activities in the control room, such as working on traditional panels and workstations, communication, and auxiliary activities. Discussions are then made on the objectives of evaluation and conditions of HSI evaluation in China nuclear industry. The corresponding experiences can be generalized to human-system interface design in other systems.

Keywords Human-system interface design · Performance assessment · Human behavior · Control room · Nuclear power plant

1 Introduction

The effect of human activities on safety is essential in safety-critical systems such as nuclear power plants, petrochemical plants, and aviation industry, etc. In socio-technical systems where people interact with technology, the reliability of

X. Dong (✉) · Y. Zhang · Y. Feng · J. Liu
Nuclear and Radiation Safety Center, Beijing, China
e-mail: dongxlchinansc@163.com; dongxiaolu@chinansc.cn

D. Pan · Z. Li
Department of Industrial Engineering, Tsinghua University, Beijing, China

Y. Xu (ed.), *Nuclear Power Plants: Innovative Technologies for Instrumentation and Control Systems*, Lecture Notes in Electrical Engineering 400,
DOI 10.1007/978-981-10-3361-2_16

human performance and the ability of human handling of abnormality are of vital importance [1]. Modern control rooms in nuclear power plants nowadays are equipped with digital instrumentation and control (I&C) systems, which results in the appearance of many task support systems like computerized procedure system, digital alarm system, and safety function surveillance system, etc. [2, 3]. Operators working in such control rooms use computerized interfaces more frequently than traditional panels, and thus with different behavioural patterns.

The primary task of operators in the control room is to monitor and control the operating status of the plant. During the normal monitoring, operators should not only pay attention to alarms that might appear but also interpret changing trends of parameters and the existing interplay between them [4]. In emergency situations, operators are required to handle the events following instructions of procedures.

Human performance can be measured by task results, which are plant conditions and task features such as time, accuracy, error, and frequency, etc. [5–7]. Form the cognitive perspective, major measures include cognitive workload and situation awareness. They can be measured based on performance, subjective rating, and physiological measurement [4, 8–10]. In tests that can validate proper integrated design, operators should be able to successfully accomplish the task with appropriate workload and certain level of situation awareness.

Considering the importance of human involvement and technological changes in human-system interaction, the influence on operators' cognitive tasks should be emphasized in the design, test, verification, and validation of interfaces [5, 6]. By observing operators' behavior and measuring corresponding performance during the test, the evaluators will be able to identify potential problems in interface design and training course development. This paper discusses the relationship between human behavior observation and HSI design along with recommendations on both issues in the realm of process control systems, especially in control rooms of NPPs. Corresponding activities include using traditional panels, using computerized workstations, communication, and other auxiliary activities.

2 Human Behavior and HSI Design

In the control room, operators' cognitive process can be affected by many factors, including external factors like task, environment, automation level, interface design, and organization, as well as internal factors such as pressure, health, and knowledge level. These factors have influence on stages of cognitive process, and thus have indications in operators' behavior and final task results. Because cognitive process cannot be directly observed, speculation can be made by observing human behavior and measuring physiological indicators like eye movement. Also, since operators will have their own understanding of the task process, methods of interview and subjective rating can be applied in the evaluation.

In engineering practice, whether human-system interface can satisfactorily support task execution should be validated through the actual task conducting process. Indirect inference on the interface design can be made by observing indications in human behavior and performance that reflect the influence of interface design on operators' cognitive process. In the analysis, these indications should be affiliated with tasks, cognitive factors, and interview results to provide a more accurate evaluation. Further recommendation on interface design, training, and organization can then be made based on the previous assessment.

In the following sections, this paper will discuss categories of operators' activities in the control room corresponding to interface design and other human factors elements. Examples of activities were observed during several field studies and validation tests. For commercially confidential reasons, the names of facilities and detailed design features are not provided in the paper.

2.1 Using Traditional Panels

In an advanced digital control room, traditional panels can be introduced as a backup for the computerized system. Operators using traditional panels can check alarms and indicators, adjust trends and displays, and implement control actions. Related activities including:

1. Moving in front of panels
   The movement of operators reveals how displays and controls are distributed on the panel. Apart from routine patrolling, frequent moves and ineffective quick returns may indicate inappropriate panel layout or unclear illustration. Walks of long distance can reveal potential problems in panel layout or task assignment.
2. Searching for information
   If operators ask for the positions of displays/controls or spend relatively long time in searching for information, possible defects in panel layout and presentation may exist. However, this situation could be improved through proper training and better communication.
3. Observing
   Awkward postures in operators' observing activities could reveal possible problems in size, presentation, position, and layout of interface elements. For example, if operators lean significantly close to the panel when examining system status, the corresponding information may not be clearly presented or well illuminated. Also, if operators turn aside or move to another spot for feedback after operation, there could be incompatibility between controls and displays.
4. Operating
   Awkward postures in operators' operating activities could reveal inappropriate deign in size, position, or control features of control devices. For example, if operators have to press a button carefully with their fingertips, it is possible that the button is too small or the distance between buttons is not enough.

2.2 Using Computerized Workstations

In advanced control rooms, operators monitor and control major plant operating functions on computerized workstations. Related operating actions include page switching, opening procedure, choosing/adjusting trends and parameters, managing alarm, and controlling of variables, etc. These actions are completed through inputs of mouse, keyboards, and/or touchscreens.

The advanced control room in a NPP is generally equipped with several small Visual Display Units (VDUs), and sometimes with Large Display Panels (LDPs) at the front. Operators in the working crew with different roles have different responsibility and focus on different displays located in the room. For instance, workstations located in the back are generally used by Unit Supervisor (US) and Shift Supervisor (SS) for monitoring. For the workstations in the front row, VDUs representing the status of primary loop systems are occupied by the Reactor Operator (RO), meanwhile Turbine Operator (TO) and sometimes Electric Operator (EO) are in charge of secondary loop systems. While RO, TO, and EO usually work in relatively restricted and well-defined areas, US and SS could walk around the MCR and supervise the crew. In order to accomplish tasks in sophisticated systems, operators should cooperate as a team and keep smooth communication.

Facing the complexity of tasks and control room coordination, experienced evaluators would identify several important and questionable actions for scrutiny. Since well-trained operators may have many information sources and perform with considerable speed during the test, it is better to take visual and audio recordings for further assessment alongside the field observation. Besides, if physiological measures like heart rate, eye movement, and EEG (Electroencephalogram) are recorded and analyzed, the integrated assessment can be more comprehensive [4, 9, 11].

During the analysis, inappropriate designs could be revealed by the performance data. Problems could be identified on interface design, navigation system design, information allocation and combination, interface elements design, and shortcuts design, etc. The following variables could be applied in identification of deficiency in design [12].

(1) Number of pages appeared/page transitions (visits)

Since display units are distributed in different positions and used by different operators under distinct circumstances, the values of these variables can vary considerably with the tasks that performed in the test. The results indicate the overall usage of workstations and displays.

Generally, more pages are required when wider range of information need to be accessed. However, too much page viewing may indicate inappropriate information allocation among pages, which means that the data provided on some pages may not be sufficient or cannot be easily found. Number of page transitions and visits will increase with the number of pages appeared, but short page stay time and revisits to same pages can significantly raise the number of transitions and visits. Besides important pages that require multiple visits, inefficient navigation could be responsible for this increase, in that when the pages in need cannot be assessed directly, visits to extra "middle" pages will become necessary.

Considering the nature of these variables, pages should be carefully examined with the consideration of tasks and scenarios if there are indications of possible deficiency. Improvement can be made by some extend of information combination. For instance, important information placed on isolated pages can be provided on major working page as supplement. The navigation system should have sufficient and effective connections among closely related pages. It is essential to carefully design the navigation system based on the results of task analysis.

(2) Number of visits to certain pages

Multiple visits to certain pages imply that such pages may be relatively important in the dynamic process. However, inappropriately designed navigation system or improper information allocation among pages may also result in high visits to certain pages. Therefore attention should be paid to ensure that these pages contain appropriate information and can be visited conveniently. By recording number of visits to each and every page appeared in the test, evaluators can identify critical pages and sometimes possible problems.

(3) Page stay time

Operators may stay on a page for many reasons, for instance, monitoring, operating, waiting for feedbacks, or leaving the page for later use. Page stay time is a simple and important index for the evaluation of digital systems and can be calculated for every page on every visit. Long stay duration on a page usually implies high need for monitoring or perhaps keyhole effects on few VDUs. Generally, well-trained operators are familiar with distributing information among many VDUs. There are also operating requirements on what to present on each workstation VDU. However, pressure of emergency could cause operators to ignore some VDUs and focus on one or two displays. In addition, durations of stays will change with the task evolvement. Thus, experts in plant operation need to make judgment on whether pages with long stay time are in need at that very moment.

For each VDU in different tasks/scenarios, there often exist one or two major pages that occupy the majority of the test time. It is reasonable given the task allocation and operating requirement. Though most visits are not very long, short stays that could not support even a fast scan can be indications of error recovery or defective navigation.

By integrating the results of page transitions and page stay time, evaluators may detect pages that are crucial for the tasks, moments when operators' attention gets narrow, and switches when wrong pages are opened. Attention should be paid on displays and navigation system in support of the operating tasks. Moreover, task reallocation, targeted training, and supportive interface features can relieve operators' workload and stress in facing transients.

(4) Cursor and keyboard activities

Mouse clicks and keyboard entries can illustrate how operators interact with the digital system. Statistics of cursor and keyboard activities could be based on using of an integrated system, a workstation, a VDU, or a certain page. Usually, mouse clicks are for page transitions. Other purposes include executing procedures, managing alarms, popping out menus, selecting from select boxes, and moving scroll bars, etc. The clicks are accompanied by cursor moving as an auxiliary action. Likewise, keyboard activities can include setting up target values and using function keys.

If there are no correct feedbacks for mouse clicks or keyboard activities, these activities are considered to be invalid. Apart from temporary system breakdown, Non-responsive clicks or key pressings could be caused by: (a) unclear indications of whether elements are clickable; (b) items too small, too narrow, too crowded, or too close to the edge to be conveniently clicked; (c) using function keys under wrong modes; (d) inputs blocked by safety interlocks; (e) improper sensitivity of the input device, etc.

For the conditions of cursor movement, long mouse moving time and frequent cursor movements could be the result of parameters, indications, or illustrations that are incompatible with the users' mental model. For example, elements with same functions appear at distinct positions on similar pages could cause confusion of the operator.

If the performance data indicates high mouse clicks per visit, low effective ratio of clicks and keystrokes, and long and frequent cursor movement, inappropriately designed interface elements may exist. The suggestion is to follow consistent design guidelines for interface elements like buttons, labels, data, pictures, and diagrams, etc. Also, the design of interface elements should adapt to users' habits. To be more specific, the categories and functions of the elements should be clear by the appearance. Also, the clickable items should be large enough and having proper space between each other to avoid wrong or ineffective clicks. Because operating modes and safety interlocks are features of the system, besides adding notifications on displays, related training on information input is also necessary.

2.3 Communication

To ensure successful completion of tasks performed in the main control room, especially proper handling of accidents, full cooperation of operators is essential. Rephrasing and verifying in communication contribute greatly to better information exchange and team performance. Therefore, operators are required to keep two-way commination in order to make the process smooth and clear. Communication language can reveal characteristics of the crew, such as skills, openness, coordination, and team spirit. Also, the style and contents of language can demonstrate the process of information gathering, problem solving, and plan execution, as well as real time pressure and workload of the crew. Observers will be able to predict the situation awareness of crew members, and to determine mendable places in training, organization, staffing, and interface design.

2.4 *Other Auxiliary Activities*

Apart from direct monitoring and operating, operators will also need to perform auxiliary activities, for example, walking to the file cabinets to fetch procedures, looking through procedures for the execution page, and connecting personnel inside and outside the plant using broadcast or telephone. Physiological measurement during the test can indicate how well facilities are placed in the room. For instance, if operators have difficulty in crossing the passage way, holding heavy procedures, or making phone calls during the test, improvement should be made on the related devices, according to the task requirement.

3 Discussion

By measuring human performance, one can assess interface design, training, and organization of the plant. The results can facilitate corresponding design of the interface and training course, which can better support operators' task performance. The analysis should synthesize related information for thorough study of the important tasks and process.

There are four levels of objectives in evaluation: (1) the results of task execution satisfy the requirement of operation and safety; (2) the execution process should be smooth and without hindrance; (3) the workload and situation awareness of operators are at appropriate levels; (4) the above objectives can be achieved in real practice, that is, can be generalized to conditions of greater scale. These four levels of objectives are gradually enhanced, which considers not only the task requirement, experts' evaluation, and design conventions, but also the variation of human performance and changes in real conditions [5, 6]. Therefore, the observation of human performance can focus on unnatural activities, error recovery, and situations of high workload on the basis of successful tasks completion. Questions on any level could imply places for improvement in areas such as interface design, training, and organization.

When conducting evaluation tests, methods of walkthrough, mock-up, and prototype with simulator can be applied in the process. Evaluators can use various kinds of tools such as human factors guidelines' checklist, questionnaire, interview, and recording analysis. Recordings on system status, task performance, human behaviour, communication, and physiological data can be taken during the evaluation tests. Among all the data collected, evaluators need to select important sets and use appropriate methods in data analysis and explanation in order to prove that the design satisfies the requirements of the above four levels. If not, some emendations should be taken place.

Serving the objective level two, the technique in identifying potential problems considering relationships between human performance assessment and HSI design summarized in this paper is an important part in the analysis and explanation process. By spotting tasks and processes that require scrutiny, the evaluators can find and solve the possible problems more effectively and efficiently. Though data recording may require hardware technology, the analysis and explanation focus more on domain knowledge, statistical knowledge, and experience in Human Factors Engineering (HFE). In China, though some attempts were made in the nuclear industry to improve interface design through performance-based testing, the majority of work is still on proving successful task completion. Sometimes, operators' subjective feedbacks were collected as a supplement in support of some small corrections. In general, the existing evaluations of NPPs' HSI in China can hardly reach the level 2: "the execution process should be smooth and without hindrance", let along higher levels. It is an urgent need that more technicians with proper trainings of human factors engineering should enter the domain of nuclear industry. At the same time, trainings on HFE should also be provided to practitioners in the industry.

4 Conclusion

In the control room of NPPs, activities involve using of traditional panels and computerized workstations, communication, and other auxiliary activities. Generally, the assessment of control room design in complex systems has common objectives in spite of great difference in task domains. The ways of interacting with panels and computers as well as methods of assessing human performance and interface design in modern systems are similar in nature. Though the conclusion is based on observations of operators' activities in the control room of nuclear power plants, the results and recommendations in this paper can be generalized to the design and evaluation of human-system interface in other systems.

References

1. Hollnagel, E. (1998) Cognitive Reliability and Error Analysis Method (CREAM). Elsevier
2. Committee on Application of Digital Instrumentation and Control Systems to Nuclear Power Plant Operations and Safety, Commission on Engineering and Technical Systems, Division on Engineering and Physical Sciences, National Research Council (1997) Digital Instrumentation and Control Systems in Nuclear Power Plants: Safety and Reliability Issues. pp. 1–31. National Academy Press, Washington, D.C.
3. Zhang, Y.F. (1997) Computerized Operator Support System for Nuclear Power Plant. Nuclear Power Engineering. 18(2), 163–169

4. Mumaw, R.J., Roth, E.M., Vicente, K.J., Burns, C.M. (2000) There Is More to Monitoring a Nuclear Power Plant than Meets the Eye. Human Factors: The Journal of the Human Factors and Ergonomics Society. 42(1), 36–55
5. O'Hara, J.M., Higgins, J.C., Persensky, J.J., Lewis, P.M., Bongarra, J.P.: NUREG-0711, Human Factors Engineering Program Review Model. Rev. 3. U.S. Nuclear Regulatory Commission, Washington, D.C. (2012)
6. O'Hara, J.M., Stubler, W.J., Higgins, C., Brown, W. (1997) NUREG/CR-6393, Integrated System Validation: Methodology and Review Criteria. U.S. Nuclear Regulatory Commission, Washington, D.C.
7. Meister, D. (1985) Behavioral Analysis and Measurement Methods. Wiley, New York
8. Gawron, V.J. (2008) Human performance, workload, and situational awareness measures handbook. Second ed. CRC press, Boca Raton
9. Tran, T.Q., Boring, R.L., Dudenhoeffer, D.D., Hallbert, B.P., Keller, M.D., Anderson, T.M. (2007) Advantages and Disadvantages of physiological assessment for next generation control room design. In IEEE 8th Human Factors and Power Plants and HPRCT 13th Annual Meeting, vol. 1, pp. 259–263. IEEE
10. Endsley, M.R., Garland, D.J. (2000) Situation Awareness: Analysis and Measurement. CRC Press
11. Ha, J.S., Seong, P.H., Lee, M.S., Hong, J.H. (2007) Development of human performance measures for human factors validation in the advanced MCR of APR-1400. IEEE transactions on nuclear science. 54(6), 2687–2700
12. Dong, X.L., Song, F., Li Z.Z., Zhang, S.S. (2013) Data extraction and analysis for integrated system validation of a nuclear power plant. Nuclear Engineering and Design. 265, 826–832

Qualification Test Standards Research of Electrical Equipment Important to Safety of Nuclear Power Plants

Yin-Hui Guo, Zhong-Qiu Wang, Yun-Bo Zhang, Yan Feng, Ning Qiao and Jing-bin Liu

Abstract Safety digital I&C system has been gradually used to the new nuclear power plant I&C system or renovation project of the nuclear power plant. Nuclear equipment reliability directly affects the safety operation of nuclear power plant, so the safety digital I&C system should be carried out Equipment qualification (EQ) in accordance with the relevant regulation and standards. EQ test is effective measure to prevent the common failure caused by operating conditions and environmental conditions. And also can meet the single failure criterion of nuclear I&C equipment. Type testing is still method of choice for the safety electrical equipment. However, in terms of safety electrical EQ, multiple international standard system lack of detailed guide. This paper analyzes the required qualification standards. Combined with the international EQ experiences then put forward requirements of qualification test items, test levels and qualification process.

Keywords Nuclear power plant · Safety · Electrical equipment · Qualification test

Y.-H. Guo · Z.-Q. Wang · Y.-B. Zhang (✉) · Y. Feng · N. Qiao · J. Liu
I&C Department, Nuclear and Radiation Safety Center, Beijing, China
e-mail: zhangyunbo@chinansc.cn

Y.-H. Guo
e-mail: gyh86126@126.com

Z.-Q. Wang
e-mail: wangzhongqiu@chinansc.cn

Y. Feng
e-mail: fengyan@chinansc.cn

N. Qiao
e-mail: qiaoningthu@foxmail.com

J. Liu
e-mail: liujingbinjob@163.com

Y. Xu (ed.), *Nuclear Power Plants: Innovative Technologies for Instrumentation and Control Systems*, Lecture Notes in Electrical Engineering 400,
DOI 10.1007/978-981-10-3361-2_17

1 Introduction

Digital instrument and control (I&C) system is the central nervous system in nuclear power plants (NPPs). The reliable and economical operation of NPP units largely depends on the performance level of I&C system. Qualification technology of class 1E equipment is necessary measure to guarantee I&C system can operate normally when it withstands normal, abnormal service conditions and accident conditions. And qualification technology of class 1E equipment is one of the key technologies to realize autonomy intellectual property rights of I&C system in NPPs.

In qualification of 1E equipment, there are a number of international standards and guides issued respectively by Institute of Electrical and Electronics Engineers (IEEE), International Electro technical Commission (IEC), International Atomic Energy Agency (IAEA), U.S. Nuclear Regulatory Commission (NRC) etc. However, among these standards for EQ are not entirely consistent. Consequently, different countries have large different requirements and actual practice in the EQ. Although part of IEEE and IEC standards are planning to unification, but still has a long way to go. On the other hand, only have the general requirements for EQ in the relevant standards, did not give the exact qualification guide. Therefore it is necessary to analyze and compare research on international standards requirements of EQ, then combined with the practical experience in the nuclear power plant safety review, considers the requirements of qualification test, finally put forward a set of suitable for nuclear safety electrical EQ test requirements.

Considering most safety electrical equipment installed outside the containment and corresponding to the RCC-E specification of K3 class equipment, this article describes EQ only for K3 class electrical equipment.

2 International Standards in EQ

2.1 Status Standard in EQ

Internationally, it is acknowledged that NRC cooperates with IEEE and IAEA cooperates with IEC so as that there have two standard architectures which can be called as American architecture and European architecture. American architecture consists of codes, regulatory guides (R.G.) and standard review plan (SRP) all issued by NRC and IEEE standards were issued by IEEE. European architecture consists of codes and guides which are issued by IAEA and IEC standards which were issued by IEC. In addition, there are several branches existing in European architecture such as in France and Germany [1].

American architecture consists of 10CFR50, R.G. and SRP NUREG 0800 and IEEE standards. 10CFR50 stands on the top of the architecture. R.G. and SRP analysis and explain codes' requirements, and endorse relevant IEEE standards to satisfy it so as to give relevant guidance for NRC staffs in review activities.

The standards above do not provide detailed guidance or advice in EQ. DI&C-ISG-06, as licensing process for digital I&C system, points out that R.G.1.209 endorses IEEE 323-2003 for mild environment qualification. R.G.1.209 considers the mild environment qualification practices endorsed in this guide are equivalent to, and consistent with those described in EPRI TR-107330 (technical report issued by Electric Power Research Institute). And also R.G.1.209 endorses R.G.1.180-2003 and EPRI TR-102323-1996 as EMC testing guide. Besides environmental qualification, EQ shall also contain seismic qualification. As seismic test guide R.G.1.100-1988, it endorses IEEE 344-1975 [2, 3].

European standard architecture is based on IAEA codes, guides, and IEC standards. In Europe, France and Germany built their own standard architecture, especially RCC-E and its supporting specifications are widely used in EQ of pressurized-water reactor NPPs in the world. RCC-E volume B provides relevant requirements about qualification and approval. And chapter B2600 points out that alternative EQ method can be used but it should conform to IEC60780. The architecture in Germany is based on KTA series which were issued by Nuclear Safety Standards Commission (KTA), and KTA 3503 is for type testing of electrical modules for the safety related instrumentation and control system. KTA3503 also endorses IEC60780 as system qualification standard. Thus IEC60780 is the main standard in EQ in European architecture from above analysis.

2.2 *Analysis of Applicable Standards in EQ*

2.2.1 Three-Level Standard Architecture in EQ

Three-level standard architecture should be used in EQ: main standards, supporting standards/guides, general standards. Main standards provide general process, procedure and method of EQ. In general, detailed test parameters cannot be found in main standards. However, supporting standards and guides will do. And most of this supporting standards and guides endorse general standards. For example, R.G.1.180 endorses IEC61000-4 standard series and IEC61000-6 standard series.

2.2.2 Main Standards

In American standard architecture, IEEE323 is the main standard for class 1E electrical EQ, while in European architecture, IEC60780 is. RCC-E volume B is the main standard in France and KTA3503 in Germany.

Comparing the content of these main standards, it can be found that the requirements of qualification items, qualification methods of RCC-E volume B and IEC60780 are equivalent. Although IEC60780 is main standards in Europe, but

RCC-E volume is widely used in Chinese EQ of in-service and in-construction NPPs. Therefore the relevant requirements of RCC-E should be referenced because autonomy intellectual property rights of digital I&C system may be used in NPPs with pressurized water reactor. On the other hand, IEEE323 is from American architecture. And American architecture is paid more and more attention because of NRC's authority in the area of nuclear power and the other I&C platforms such as Common Q platform have been all qualified according to it. So IEEE 323 should also be referenced.

2.2.3 Supporting Standards, Guides and General Standards

IEEE323, IEC60780, and RCC-E all point out that type test can be divided into different phases and in a specified order. Different phases should conform to different supporting standards and guides. Comparing these main standards, it can be concluded that qualification test can be according to the following order: reference test, extreme environmental test, aging test, accident and post-accident function test. The following sections will analyze applicable supporting standards, guides and general standards of these four phases.

2.2.4 Conclusion of Applicable Standard Analysis

From the analysis of applicable supporting standard, guides and general standards in four different phases, it can be concluded that the applicable standards which have three levels for digital I&C system EQ as shown in Fig. 1.

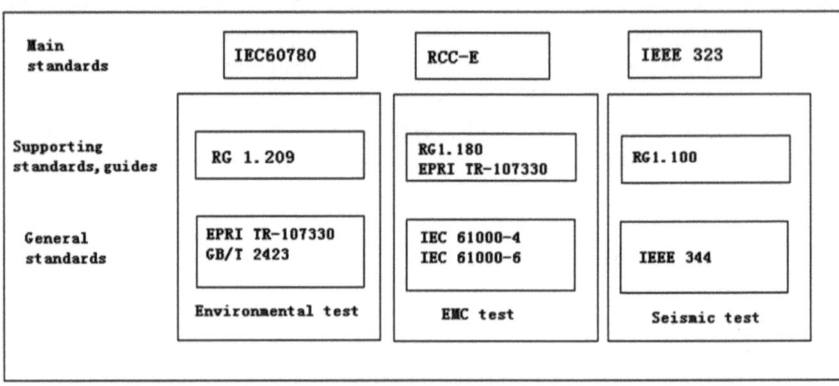

Fig. 1 Standard architecture applicable to EQ

3 Test Item of EQ

3.1 Test Item Overview

IEC 60780 point out EQ type tests shall be divided into three main test groups.

- Group 1: To check the functional characteristics under normal ambient conditions and in all specified limits of normal operation.
- Group 2: To demonstrate the seismic resistance of equipment.
- Group 3: To demonstrate the resistance of the equipment against accident and post-accident conditions.

This article focus on the K3 equipment is not affected by environmental conditions after accident and post-accident. Therefore do not consider the Group 3 requirements [4].

In the chapter 6.3.1.7 of IEEE 323 are mentioned test sequence should be follow as: Specified baseline functional tests shall be performed under normal conditions; The test sample shall be operated to the extremes of all performance, operating, surge voltages, and electrical characteristics given in the equipment specifications, excluding design basis event and post-design basis event conditions, unless these data are available from other tests (e.g., design verification tests) on identical or similar equipment; When required, the test sample shall be age conditioned to simulate its functional capability at the end of its qualified life. The test sample shall be subjected to specified non-seismic mechanical vibration; The test sample shall be subjected to simulated OBE and safe shutdown earthquake (SSE) seismic vibration in accordance with IEEE 344 [5].

RCC-E requires to ensure that equipment installed outside the containment is capable of performing its specified functions under normal ambient conditions and under seismic loading, as well as accidental ambient conditions specified for certain items of equipment [6].

Integrated the above requirements of test contents and test process, EQ should include the following stages:

- Reference test.
- Extreme environmental test.
- Aging test.
- Accident and post-accident function test.

3.2 Reference Test

The objective of this test is to verify the functional characteristic of the equipment under normal environmental condition. This test includes (1) electrical interface characteristics tests and measurement of insulation resistance, (2) functional characteristics measurement. The functioning of the equipment as specified in the data sheet shall be demonstrated.

3.3 Extreme Environmental Test

During this phase the equipment shall be operated to the extreme limits of its utilization field (electrical and environmental) indicated in its performance specification. Safety electrical equipment should be checked: the specified limits of normal supply voltage (or frequency) field; the extreme limits of the temperature range; electrical or electromagnetic disturbances, conducted and/or radiated.

Extreme test for power supply, AC-powered devices are required function performance under supply voltage and frequency limit. The choice of extreme power requirements should meet the specifications and need to combine the power supply and frequency standards of each country.

Climatic test shall be demonstrated may be subjected during normal operation limits of temperature and humidity. Also taking into account the extreme conditions when ventilation heating equipment failure. EPRI TR-107330 provides test temperature and humidity curve in Fig. 2 [7].

Figure 2 give fully consideration to the operating equipment may be subjected to the high temperature and humidity, low temperature and humidity. Test conditions may cover most of the operating condition of electrical equipment.

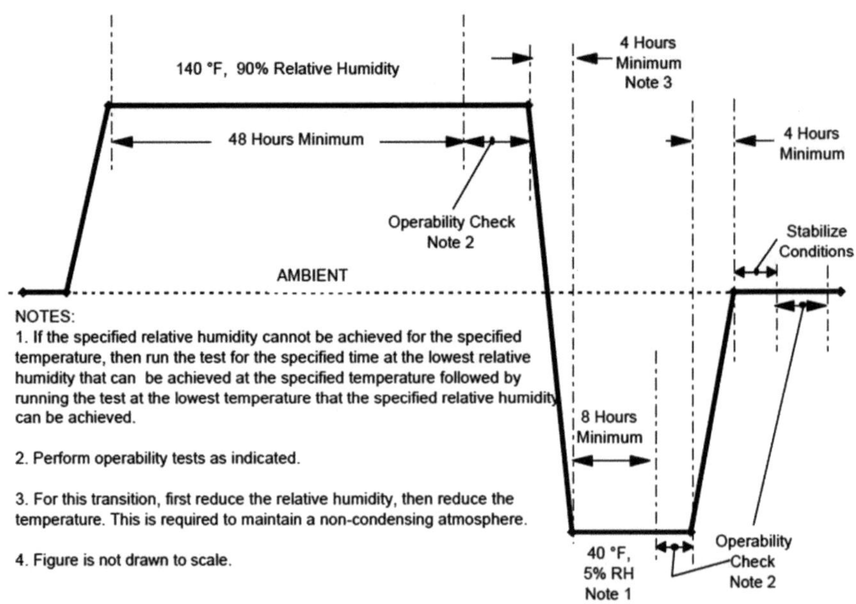

Fig. 2 Temperature and humidity requirements of EPRI 107330

About EMC test, R.G.1.209 states that R.G.1.180 and EPRI TR-102323 provide guidance for EMC test. R.G.1.180 and TR-102323 endorse general standard MIL-STD, IEC61000-4 series and IEC61000-6 series. R.G.1.180 gives test grades and acceptance criteria of radiation emission, radiation immunity, and conduction emission and conduction immunity [8, 9].

For operating, the equipment surrounds the vibration source. According to the vibration amplitude and acceleration, the equipment should be tested the ability to withstand vibration. Test methods can be followed the requirements of IEC 60068-2-6.

3.4 Aging Test

The assessment of equipment aging effects in connection with a type test program is required to determine if aging has a significant effect on operability. Requirements of ageing test in IEC60780 and RCC-E are equivalent. Based on the characteristics of the electrical equipment to choice the following test:

- Thermal test and/or thermal tests with mechanical effects.
- Mechanical vibration tests.
- Prolonged operating test.
- Irradiation ageing test.

3.5 Accident and Post-accident Function Test

Accident and post-accident function test will be performed to verify the functionality of the equipment under accident condition. In general, only seismic condition should be considered because most safety class electrical equipment will be installed outside the containment.

According to IEEE 344, if equipment will be installed on the specified floor location and the floor response spectrum is available, then the seismic test should use artificial time history method. On the contrary, if the equipment installation location or the response spectrum is not available, or the equipment will be installed in different location, then it is better to use the single frequency sine beat method to simulate the seismic condition [10].

For time history method, the test should be performed on a dual-axis or tri-axis shake table. And the table shall output multi-frequency.

Single frequency wave method means the single-axis shake table shall output specified single frequency wave from 1 to 33 Hz which is the seismic frequency band. Test frequency points shall be selected based on the center of enlarged area, and then the frequency points shall be extended to both ends with the one-third octave. Each sine wave beat should include 5–10 sine wave. And one time OBE or

SSE should include 5 or above beats. Time interval should be more than 2 s to avoid equipment overlap effect. IEC 60980 gives the magnification factor from the input time domain acceleration to test spectrum based on the equipment damp ratio, the number of beat and the number of sine wave of each beat. Therefore if test spectrum is available, then the input acceleration level can be calculated according to the magnification factor.

In general, seismic test will perform 5 OBEs followed by 1 SSE.

4 Process Assurance in Qualification Test

4.1 Preparation of Test Specimen

For the condition that qualification equipment is combined by multi-cabinets or has complex configuration, a qualification test specimen (referred to as the specimen) which is enough to represent the qualification equipment should be prepared. The representative of the specimen is reflected in following aspects:

- Representative of function. The specimen should have the typical function of qualification equipment, including typical input and output, operation process, data communication, typical isolation, typical information display, and typical monitoring and diagnostic function.
- Representative of products type. For the modules which are used to realize the hardware, software, man-machine interface, and data communication functions and the auxiliary modules, at least one of each type of module should be configured in test specimen.
- Representative of mechanical structure. The dynamic structural characteristics of specimen should be similar with qualification equipment.
- Representative of application configurations. The specimen should be configured similar with or more rigorous than the qualification equipment, including the cabinet temperature rise and interface connections in actual operation.

4.2 Preparation of Test Instruments

The test equipment under strict verification and evaluation and certified third party lab are effective methods to ensure the credibility of the test data. Test equipment and third party lab should be chosen according to the characteristics of the specimen before qualification test, the requirements including:

- The generic instruments and meters which are used in the test must be within calibration period validity. The calibration records should be contained in test reports according to IEC60780-1998 section 6.3.

- Some auto-test or auto-monitor equipment which is self-developed shall have the developing process documents and verification record which can be traceable.

4.3 *Evaluation of the Third-Party Test Labs*

Test Lab satisfied with requirements of test item shall be selected for qualification test. The organization responsible for qualification should evaluate the third party lab, includes: authorization, parameters of test equipment, capability of test engineers and quality assurance.

5 Conclusions

Safety functions of the safety system are guaranteed by safety class electrical equipment and the qualification of electrical equipment become the important approach of deep defense. Only if the equipment undergoes the strict test can guarantee the equipment can operate reliably under different abnormal environmental conditions.

This paper analyzed the requirements of related standards of qualification, concluded the requirements of qualification test of electrical equipment combined with qualification test experience of other international DCS platform. The conclusion of this paper will be reference for persons of I&C electrical EQ.

References

1. HUANG Weijie, (2014) Nuclear Power Engineering. A Preliminary Study on Qualification of Instrumentation and Control System for Nuclear Power Plants. Vol. 35. No. 6
2. R.G.1.100-1988 Seismic Qualification of Electric and Mechanical Equipment for Nuclear Power Plants
3. EPRI TR102323-1997 Guidelines for Electromagnetic Interference Testing in Power Plants
4. IEC 60780-1998, Nuclear power plants - Electrical equipment of the safety system - Qualification
5. IEEE Std. 323-2003, IEEE Standard for Qualifying Class 1E Equipment for Nuclear Power Generating Stations
6. RCC-E-2005, Design And Construction Rules for Electrical Equipment of Nuclear Islands
7. EPRI TR107330-1996 Generic Requirements Specification for Qualifying a Commercial Available PLC for Safety-Related Application in Nuclear Power Plants
8. R.G.1.209-2007 Guidelines for Environmental Qualification of Safety-Related Computer-Based Instrumentation and Control Systems in Nuclear Power Plants
9. R.G.1.180-2003 Guidelines for Evaluating Electromagnetic and Radio-Frequency Interference in Safety-Related Instrumentation and Control Systems
10. IEEE 344 2004 IEEE recommended practice for seismic qualification of class 1E equipment for nuclear power generating stations

The Several Issues in Safety Review of Digital Control System in Chinese Nuclear Power Plant

Yun-Bo Zhang, Zhong-Qiu Wang, Yan Feng, Jing-Bin Liu and Yin-Hui Guo

Abstract With the extensive application of digital control system (DCS) in Chinese nuclear power plant, the safety of DCS attracts people's attention. First of all, this paper briefly introduces the overall structure of DCS. Then it summarizes the several key issues about DCS from the perspective of nuclear safety review: the single failure criterion, testability, diversity, software verification and validation (V&V), configuration management, which are common issues in the safety review. These issues are analyzed according to the relevant regulation and standards. At last, some solutions are given to these common issues, and some suggestions are made for future review.

Keywords Nuclear power plant · DCS · Safety review

1 Introduction

Digital control system is one of the most important systems in nuclear power plant. DCS is more like the "central nervous system" of the entire nuclear power plant, and its stability is the key for the safety, reliability and economic of operation. At present, DCS technology is or will be used in most of Chinese nuclear power plants.

Y.-B. Zhang · Z.-Q. Wang · Y. Feng (✉) · J.-B. Liu · Y.-H. Guo
I&C Department, Nuclear and Radiation Safety Center, Beijing, China
e-mail: fengyan@chinansc.cn

Y.-B. Zhang
e-mail: zhangyunbo@chinansc.cn

Z.-Q. Wang
e-mail: wangzhongqiu@chinansc.cn

J.-B. Liu
e-mail: liujingbinjob@163.com

Y.-H. Guo
e-mail: Gyh86126@126.com

Y. Xu (ed.), *Nuclear Power Plants: Innovative Technologies for Instrumentation and Control Systems*, Lecture Notes in Electrical Engineering 400,
DOI 10.1007/978-981-10-3361-2_18

Compared with traditional analog technology, the application of DCS can improve the efficiency of the nuclear power plant, safety and reliability. With the application of DCS in nuclear power plant, the related operation event leads to widespread concern. Therefore, the review of the DCS is one of the key issues in nuclear safety review. This paper analyzes several important issues in safety review of DCS, and gives some suggestions for future review.

2 The Structure of DCS in Nuclear Power Plant

In general, DCS can be divided into 4 levels by functions: field level, automation level (individual control and measurement level), communication level, process information and control level. The field level is mainly used for detecting the parameters of process equipment, controlling the technical process according to the command, providing/controlling processing functions such as power supply equipment. The automation level is mainly used for data acquisition, signal pre-processing, logic processing, operation of control algorithm, and other functions. The communication level is mainly used for data and signal communications. The process information and control level is mainly used for information support, diagnostic, operator action and process information records, controlling unit by operating equipment and other tasks.

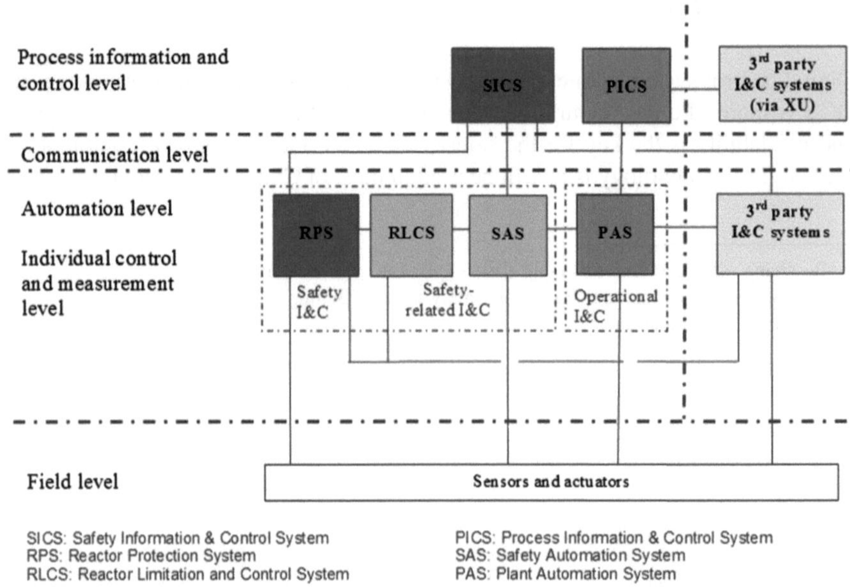

Fig. 1 DCS structure diagram

Figure 1 is the overall DCS structure diagram. Automation level, communication level, process information and control level are the focus of attention in safety review.

3 The Review Concerns

3.1 The Structure of Reactor Protection System

Reactor protection system mainly includes reactor trip (RT) system and engineered safety feature (ESF) system [1]. The RT system is made up of process measurement channel, nuclear measuring channel, acquisition and processing logic channel, reactor trip breaker, manual driving circuit and so on.

Except for adopting four redundant channels, the RT system should also adopt the principle of diversity; the protection parameters with functional diversity are assigned in different processors in order to reduce the impact of common cause failure. When a protection channel fails, if the rest of the protection channels can't meet the single failure criterion, the system will generate scram signal to achieve safe shutdown.

The ESF system consists of measurement/signal processing and logic part.

The structure of measurement/signal processing part is similar to RT system, including four redundant signal processing channels.

The logic part includes two redundant series. It receives the input signal from the signal processing channels and performs the required logic to drive the engineered safety features. Each logic series can drive minimum number of engineered safety facilities and equipment as the safety functions required. To ensure that, single failure in any redundant series will not result in the loss of protection functions.

3.2 Testability

Reactor protection system should have the ability of fault detection and testability during the reactor shutdown mode and power operation mode. The test should be conducted in several stages and each stage should be overlapped to ensure the integrity of the system test [2]. The fault detection and test of RT system should include T1 test-measuring instrument channel test, T2 test-processing unit test (digital protection system test) and T3 test-the output signal and the actuator test.

In the case of self-check failed to cover any fault, there should be specific test equipment to determine the periodic test cycle on the basis of reliability analysis.

3.3 The Diversity of ATWT and RT System

As the back-up of emergency shutdown system, anticipated transients without trip (ATWT) mitigation system should follow the principle of diversity and independence in the system design [3]. It uses the different devices (including software and hardware) to perform the functions same as RT system. For example, the functions of ATWT are implemented by the NC platform and the functions of RT are implemented by the safety platform. The diversity between different platforms are in several aspects such as design, equipment, software, human factors and so on [4].

3.4 Software Verification and Validation

In case of one nuclear power plant, its safety platform TRICON (V10.2.1) has passed the V&V conducted by the vendor itself and the independent third party. Meanwhile in the final safety evaluation report of TRICON, the NRC states that the development process of TRICON (V10.5.1) and the new/updated version of the software components meet the requirement of SRP section 7, BTP7-14, BTP7-18, EPRI TR-107330 and EPRI TR-106439, and the software modules in TRICON (V9.5.3) are also accepted due to the safety evaluation of TRICON (V9.5.3). The review conclusion requires a detailed description of the hardware and software changes from TRICON (V10.5.1) to TRICON (V10.5.3) which is used in this nuclear power plant.

For the specific TRICON application software of this nuclear power plant, the related V&V work is conducted by the independent third party. So it can effectively guarantee the independence of V&V work in management, organization and financial [5].

3.5 Configuration Management

The configuration management of DCS is used to control hardware equipment, software, and file version so as to identify the system version, help to the execution of the change, and to protect the history of configuration object and so on [6].

The following problems often exists in the configuration management work of DCS in nuclear power plant: there is no clear distinction between safety and non-safety items in the configuration state statistics; it lacks some contents such as software tool which is required in configuration management program; the naming rule of software version is unclear; the configuration state statistics can't reflect the actual situation of DCS, and so on.

A detailed and specific configuration state statistics, the naming rule of the version and the clear expression of the software version in the V&V report are helpful to ensure the implementation of the configuration management program of DCS. The configuration management can effectively control the hardware and software version in factory test phase and site commissioning phase, and it is also helpful to the following subsequent change management and update work.

4 Conclusion and Suggestion

The structure design, testability, diversity, software V&V, configuration management of DCS should meet the requirements of relevant laws and regulations.

For future review, some suggestions should be noticed. The software configuration management of hardware and software should be enhanced and improved to ensure the effectiveness; the safety software V&V activities should be carried out strictly in the operation and maintenance phase according to the requirements of relevant laws and regulations.

In recent years, some operator workstation disable events had showed up in some nuclear power plants, and the operators had to move to the BUP panel to wait for workstation restarting. These events brought some adverse effect for the stable operation of the power plant. So it should be paid more attention to the commissioning test and operation of non-safety DCS.

Acknowledgements This work is supported by the soft subject "Research and standard development of cyber security about digital I&C system in nuclear power plant". Many people have offered me valuable help in this work. I thank professor Zhong-Qiu WANG for helpful conversation, Engineer Jing-bin Liu and Yin-hui Guo for proposed changes.

References

1. ZHENG Weizhi (2012) The Fault Analysis and Application of Reactor Protection System in Nuclear Power Plant. Nuclear Electronics & Detection Technology, Vol. 32. No. 3: 337–341.
2. GB/T5204-1994, Periodic tests and monitoring of the safety system of nuclear power plant.
3. US NRC. NUREG/CR6303-1994, Method for Performing Diversity and Defense-in-Depth Analyses of Reactor Protection Systems.
4. ZHANG Yunbo (2014) Analysis of Diversity and Independence for ATWT Mitigation System in Nuclear Power Plant. Nuclear Power Engineering, Vol. 35. No. 6: 77–79.
5. IEEE Std.1012-2012, IEEE Standard for System and Software Verification and Validation.
6. IEEE Std.828-2005, IEEE Standard for Software Configuration Management Plans.

The Situation and Suggestion of Diversity Actuation System Applied in China

Jing-Bin Liu, Zhong-Qiu Wang, Yun-Bo Zhang, Yan Feng, Yin-Hui Guo and Xiao-Lu Dong

Abstract Digital control system has been widely used in nuclear power plant. It brings some obvious advantages in system design and application. Meanwhile the accompanying common cause failures (CCFs) problem becomes one of the most concerned issues. To deal with this problem, many regulations and methods are proposed. NUREG 6303 describes a method for analyzing computer-based nuclear reactor protection systems to discover design vulnerabilities with common-mode failure (US NRC. NUREG/CR6303-1994 in Method for Performing Diversity and Defense-in-Depth Analyses of Reactor Protection Systems, [1]). NUREG 7007 considers that diversity in a safety system is needed for mitigating the consequences of potential CCFs and provides guidance to the staff and nuclear industry for evaluation (US NRC. NUREG/CR7007-2010 in Diversity Strategies for Nuclear Power Plant Instrumentation and Control Systems, [2]). In this paper, the diversity actuation system (DAS) has been described in detail and the functions of DAS system have been introduced. Then several representative different reactor type in China is selected (including M310 model, Second generation plus based on M310 model, AP1000 model and EPR model), and made some comparison between these

J.-B. Liu · Z.-Q. Wang · Y.-B. Zhang · Y. Feng (✉) · Y.-H. Guo · X.-L. Dong
I&C Department, Nuclear and Radiation Safety Center, Beijing, China
e-mail: fengyan@chinansc.cn

J.-B. Liu
e-mail: liujingbinjob@163.com

Z.-Q. Wang
e-mail: wangzhongqiu@chinansc.cn

Y.-B. Zhang
e-mail: zhangyunbo@chinansc.cn

Y.-H. Guo
e-mail: guoyinhui@chinansc.cn

X.-L. Dong
e-mail: dongxiaolu@chinansc.cn

Y. Xu (ed.), *Nuclear Power Plants: Innovative Technologies for Instrumentation and Control Systems*, Lecture Notes in Electrical Engineering 400,
DOI 10.1007/978-981-10-3361-2_19

DAS systems. At the end of this paper some suggestions are proposed to the third generation nuclear power technology in China to reduce the common cause failures and enhance the reliability.

Keywords Diversity actuation system · Common-cause failure · I&C system

1 Introduction

Currently, digital control system has been used more widely in nuclear power plants (NPPs) in CHINA and abroad. Its openness, high reliability, rapidity and operability have been gradually recognized by the industry. However, due to the application of digital technology, common cause failures become one of the factors must be considered in the design [3]. The work of diversity and common cause failures is still in research stage in our country, there are no relative specific regulations and standards. But the regulator's position is very clear: HAD 102/14 [4] points out that the possibility of common cause failures of safety-critical items must be considered; it determined where the diversity, redundancy and independence should be applied to achieve the required reliability. HAD 102/16 [5] emphasizes that software CCFs is a key issue and using the diversity strategy can reduce potential CCFs effectively and improve its reliability. CCFs problems become one of the concerned issues in safety review of digital control system design and modification [6].

Diversity actuation system provides the necessary means to alleviate the consequence when design basis accidents occur due to the CCFs of digital reactor protection system. Compared with the protection and control systems it could be based on a different platform and use diverse system design, functions driven and parameter display to provide a back-up.

2 The Functions of DAS System

Although different nuclear plants types, manufacturers and technologies of DAS system may be very different in design and implementation, the main functions of DAS system are basically the same:

- Provide diverse and backup automatic drive signals such as reactor trip (RTs) and engineered safety features (ESFs) functions. When the specified parameters exceed the setting value, automatically shutdown or drive critical ESF functions to assure the integrity of the fuel cladding, primary circuit pressure boundary and containment pressure boundary;

- Provide diverse and backup manual RT and ESF functions to assure the integrity of the fuel cladding, primary circuit pressure boundary and containment pressure boundary;
- Provide adequate information of the device status and parameters on the panel station in the main control room (such as reactivity, core residual heat removal condition, primary system coolant outlet temperature, primary circuit pressure boundary condition, Containment pressure boundary etc.) to monitor and display the safety-critical parameters.

To achieve the above functions, DAS system generally has the following devices, as described in Fig. 1:

- DAS Panel station. This Panel station includes the manual operation devices, indicating instruments and alarm devices. It is mainly used for safety-critical manual functions, provides safety-critical parameters, alarm information and other display functions.
- DAS Processor Cabinets. The cabinet is mainly used for collecting, processing and outputting the signals. In order to prevent equipment malfunction and rejection, it should be set up with architectural redundancy. Each redundant subsystem would collect signals and process, compare with the set values then generate the "local signal" for voting logic.
- Anticipated Transients without Trip (ATWT) mitigation system. ATWT event is caused by control rods that can not be inserted into the reactor core due to the CCFs and the unit failed to achieve the expected transient scram. So ATWT system uses diverse equipment to achieve reactor scram. At present most of the NPPs integrate this function into DAS systems.

In order to achieve the above functions, different instrumentation and control system uses a variety of strategies to achieve.

Fig. 1 The function diagram of DAS

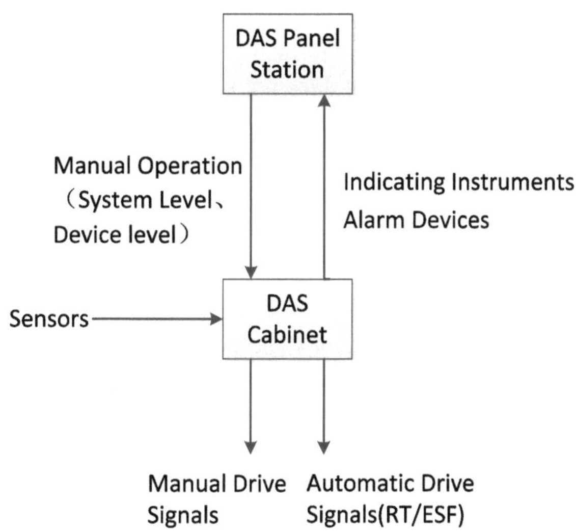

3 The Introduction of Different Diversity Methods Used in China

3.1 M310 Model

As second-generation nuclear power technology, it is widely used in CHINA's nuclear power project, and achieved good operating performances. At present, there are still a large number of operating units. The digital technology in the nuclear power field had not been universal when this type of plants completed its design, besides consensus on diversity and common cause failures had not been agreed, therefore such models did not set up a separate DAS system.

However, to avoid common cause failures since the control rods can not be inserted into the core and the unit failed to achieve the expected transient scram, it sets ATWT mitigation system [7]. The system is mainly to monitor the SG water flow and nuclear power level. When the SG water flow is lower than the setting value and the nuclear power exceed the setting power, then the auxiliary feedwater system will be activated, the turbine will be triggered and the shut-down signal will be sent out. ATWT system uses diverse equipment to achieve reactor scram, follow the principles of diversity and independence. However because the functions are simple and there are no diverse safety-critical parameters and alarm information displayed in the main control room, in subsequent improved and new designed power plants this system is integrated to the DAS system.

3.2 Second Generation Plus Based on M310 Model

DAS system of Fangjiashan/Fuqing project is different from the safety-class digital I&C platform TRICON, it uses a non-safety digital I&C platform I/A. There are enough physical and electrical isolation between DAS system and reactor protection system. DAS system does not receive the signals processed by reactor protection system, meanwhile does not send signals to reactor protection system. As stated earlier, it integrated the functions of ATWT systems. In addition, parts of the automatic functions of reactor protection system are also selected as below [8]:

- Automatic reactor trip and turbine trip;
- Main steam(-piping) system isolation actuation;
- Safety injection actuation.

There are five reactor shutdown signals, respectively: High power range neutron fluence rate, high pressurizer pressure, low pressurizer pressure, 2 loops of low reactor coolant flow rate, low pressurizer pressure (this signal will trigger safety injection at the same time).

DAS system is not provided with system-level and device-level manual operations as well as safety-critical parameters and alarm information display. Backup manual control and display functions are mainly achieved through safety video display unit (SVDU), network computer-video display unit (NC-VDU) and back-up disk.

As a diversity of design, the reactor trip of automatic functions is accomplished through cutting rod position indicating and rod control system (RGL) power supply, which is different from the way of protection system by opening the trip breakers. The ESF functions of DAS system are accomplished through its own cabinet which bypassing the logical processing of reactor protection system. The signals of ESF is sending to the priority logic processing module (PLM), and then sending to the relevant actuator. Furthermore, the output signals of DAS systems are initiated by energized way not de-energized way like protection system.

3.3 AP1000 Model

As the third-generation nuclear power technology AP1000 is imported from Westinghouse. Its safety grade platform is COMMON-Q to perform the functions of the reactor protection system and its non-safety grade platform is OVATION to achieve majority control functions of nuclear island/conventional island/BOP. The safety system (PMS) uses hardware and software that is based on microprocessor and Plant Control System (PLS) is also microprocessor-based. In order to achieve diversity, the DAS system uses ALS platform that is based on Field Programmable Gate Array (FPGA).

DAS system selects parts of the automatic functions of the reactor protection system [9]:

- Automatic reactor and turbine trip;
- Automatic CMT (coolant makeup tank) valve actuation and RCP (reactor coolant pump) trip;
- Automatic PRHR (passive residual heat removal) discharge valve actuation and IRWST (in-containment refueling water storage tank) gutter isolation valve actuation;
- Automatic containment isolation valve actuation and PCS (passive containment cooling system) valve actuation.

DAS system uses independent sensors compared with PMS system, such as hot leg RTDs, containment RTDs, core exit thermocouples, steam generator level transmitters, etc.

The trip signal of automatic functions drives the generator set trip, and then CRDM (control-rod device mechanism) losing of power and dropping of control rod. This method is different from the manner of PMS that uses trip breaker to shutdown. ESF functions also uses diverse drive interface compared with PMS, and these devices can operate independently without affecting each other.

In addition, there are system-level manual operation functions in DAS Panel station and processor cabinets (partly). Its output signals are connected via hard-wired to the final load, thus can completely bypass the PMS and DAS automatic drive logic path. To support manual actuation, there are adequate instrument indications and alarms provided in the DAS Panel station and processor cabinets.

3.4 EPR Model

EPR is the third-generation nuclear power technology imported from AREVA. Due to the different strategies of diversity, EPR does not actually set up a separate DAS system strictly. It uses defense in depth strategy and adds a hard core system (HKS) to maintain the diversity. Its safety grade platform is TXS to perform the functions of the reactor protection system and its non-safety grade platform is SPPA-T2000 to achieve control functions [10].

The I&C system includes 2 levels and is made up of 9 subsystems. Level 1 layer contains process automation system (PAS), safety automation system (SAS), that are based on SPPA-T2000 platform; reactor protection system (PS), reactor control surveillance and limitation system (RCSL), severe accident I&C system (SA I&C), priority actuation and control system (PACS), hard kernel system (HKS), that are based on TXS platform. Level 2 layer contains process information and control system (PICS) and safety information and control system (SICS).

Compared with other power plant, there are a large number of F1B safety functions implemented in SPPA-T2000 SAS system, and post-accident related functions are implemented in SPPA-T2000 SAS system. Since the SPPA-T2000 is the lower safety class platform, in the event of design basis accident with SPPA-T2000 platform failure, some functions that bring the power plant into safe shutdown state will fail and lose. Therefore, the HKS functions are added into the design of I&C systems and the DEC-B functions of SAS subsystem are allocated to the dedicated cabinet. HKS system is used for the event of design basis accident with SPPA-T2000 platform failure. SAS DEC-B and SA I&C are used for DEC-B events together.

3.5 Summary

Table 1 is the situation of digital I&C system applied in CHINA. This table introduces the NPP's platform, supplier, non-1E platform and DAS/ATWT situation. It contains five types of NPP and the DAS system of Sects. 3.1–3.4 is included.

Table 1 Digital I&C system situation applied in China

NPP	1E platform	Supplier	Non-1E platform	DAS/ATWT situation
Tianwan1,2	TXS	AREVA	TXP	Only ATWT system (Seismic)
Lingao3,4	TXS	AREVA	TXP	
Tianwan3,4	TXS	AREVA	SPPA-T2000	
Taishan1,2	TXS	AREVA	SPPA-T2000	ATWT+HKS, same with Sect. 3.4 introduction (Seismic)
Hongyanhe1–4	Meltac	Mitsubishi	Hollias	Only ATWT system (Seismic)
Ningde1–4	Meltac	Mitsubishi	Hollias	
Yangjiang1–4	Meltac	Mitsubishi	Hollias	
Fangchenggang1,2	Meltac	Mitsubishi	Hollias	
Fuqing1–4	TRICON	INVENSYS	FOXBORO I/A	DAS system is based on non-1E platform, same with Sect. 3.2 introduction (Seismic)
Fangjiashan1,2	TRICON	INVENSYS	FOXBORO I/A	
Changjiang1,2	TRICON	INVENSYS	FOXBORO I/A	
Sanmen1,2	Common Q	Westinghouse	Ovation	DAS system is based on FPGA technology, same with Sect. 3.3 introduction (Non-seismic)
Haiyang1,2	Common Q	Westinghouse	Ovation	
Yangjiang5,6	FirmSys	CTEC	HOLLiAS-N	DAS system is based on diverse platform compared with protection system. It is like Sect. 3.2, but added some functions such as system-level and device-level manual functions, instrument indications and alarms (the final solution is not confirmed yet) (Seismic)

4 Conclusions

We can make a thinking and conclusion from the introduction and comparison of the different DAS system.

The DAS system of AP1000 has the most comprehensive and deepest diverse means in design. It uses a different I&C platform compared with protection system (especially different from control system and its digital technology is based on FPGA). Besides it uses a different shutdown strategy and has part of automatic functions of the reactor protection system and manual functions. There are adequate instrument indications and alarms provided in the DAS Panel station and processor cabinets. DAS system also uses a different dedicated driver interface and sensor. The only drawback is that the DAS system is not designed to be ant seismic, just not caused an inadvertent actuation of a squib valve.

M310 and EPR model does not set up a separate DAS system strictly. Though EPR adds HKS system to ensure safe shutdown when the event of design basis accident with SPPA-T2000 platform failure. However it does not set up automatic functions in response of design basis accidents with failure of TXS safety platform. There is still much room for improvement.

M310 plus model takes many effective measures in diversity design such as automatic trip and ESF functions of DAS system. But it does not contain system-level and device-level manual functions, independent parameters display, sensors and driver interfaces are shared with protection system, so there is also much room for improvement. Table 1 has shown that subsequent Yangjiang5,6 and Hongyanhe5,6 units have some improvements in these aspects, but the final solution is not determined.

HPR1000 (Hua-long Pressurized Reactor) project has reached the international advanced level of third generation nuclear power technology in safety indicators and technical performance. Its instrumentation and control system design can draw on the above diversity strategy of DAS system, enhance advantages and avoid disadvantages, to implement the principle of defense in depth and diversity.

Acknowledgements This work is supported by the soft subject "Research and standard development of cyber security about digital I&C system in nuclear power plant". Many people have offered me valuable help in this work. I thank professor Zhong-Qiu WANG for helpful conversation, Senior Engineer Yun-Bo ZHANG for assistance with the information material.

References

1. US NRC. NUREG/CR6303-1994, Method for Performing Diversity and Defense-in-Depth Analyses of Reactor Protection Systems.
2. US NRC. NUREG/CR7007-2010, Diversity Strategies for Nuclear Power Plant Instrumentation and Control Systems.
3. US NRC. NUREG 0800 Chapter 7-2007, Guidance for evaluation of diversity and defense in-depth in digital computer-based instrumentation and control systems.
4. HAD 102/14-1988, Safety related instrument and control system in nuclear power plant.
5. HAD 102/16-2004, The important safety system based on computer software in nuclear power plant.
6. Mao Congji, Wu qi (2012) Discussing the Design of Digitized I&C for NPP from the Perspective of Safety Review. PROCESS AUTOMATION INSTRUMENTATION, vol. 7:39–42.
7. Zhang Yunbo, Zhang Mi, Huang Weijie, Mao Congji, Li Shixin, Yin Baojuan (2014) Analysis of Diversity and Independence for ATWT Mitigation System in Nuclear Power Plant. Nuclear Power Engineering, Vol. 35. No. 6:77–79.

8. Xiao Peng, Liu Hongchun, Zhou Jixiang, Guan Zhonghua (2014) Design of Diverse Actuation System in Nuclear Power Plant. Nuclear Power Engineering, Vol. 35. No. 2:90–93.
9. YU Jinbo (2008) The diversity analysis of AP1000 drive system. Heilongjiang science and technology information, Vol. 33:53–54.
10. JIN Si-qi, PENG Jin, PENG Hua-qing, ZHOU Wei-hua (2012) Research on the strategy coping with DCS platform failure of EPR. Nuclear Science and Engineering, Vol. 32 s2:105–110.

The Application of the Wireless Sensor Network in Intelligent Monitoring of Nuclear Power Plants

Jianghai Li, Xiaojing Kang, Zhenhai Long, Jia Meng and Xiaojin Huang

Abstract The wireless sensor network (WSN) has great potential in monitoring equipment and processes of nuclear power plants (NPPs). The WSN can not only lower the cost of regular monitoring, but also enable the capability to achieve intelligent monitoring. The massive and heterogeneous monitoring data collected by the WSN can contribute various monitoring applications, including the cyber and physical security defense, the fault detection and diagnosis, and the advanced operation and maintenance. The Wi-Fi technology is promising to be the underlying platform for the WSN in NPPs. Specific data-driven statistical algorithms for anomaly detection and identification are demonstrated.

Keywords Nuclear power plants · Wireless sensor networks · Intelligent monitoring · Cyber security · Fault diagnosis · Nuclear security

1 Introduction

The sensor network based on wireless communication technology has been widely used in industry, agriculture, military, transportation and other fields, but not yet in nuclear power plants (NPPs). In China, the wireless system in NPPs is used merely for voice communication, e.g. the wireless telephone system in the AP1000 NPP designed by Westinghouse. Around the world, a few NPPs apply the wireless technology to data transmission, equipment monitoring, radiation measurement, personnel positioning, fire control and video surveillance [1]. The advantages of the wireless communication technology are not fully played.

J. Li (✉) · X. Huang
Key Laboratory of Advanced Reactor Engineering and Safety of Ministry of Education, Collaborative Innovation Center of Advanced Nuclear Energy Technology, Institute of Nuclear and New Energy Technology of Tsinghua University, Energy Science Building, Tsinghua University, Beijing, China
e-mail: lijianghai@tsinghua.edu.cn

X. Kang · Z. Long · J. Meng
Chinergy Co., Ltd, Beijing, China

Y. Xu (ed.), *Nuclear Power Plants: Innovative Technologies for Instrumentation and Control Systems*, Lecture Notes in Electrical Engineering 400,
DOI 10.1007/978-981-10-3361-2_20

At present, the process control system of NPPs is mainly composed of wired connection between instruments and the Distributed Control System (DCS). Instrument signals and control commands are transmitted by hard wires. Although the wired connection is assumed to be safer and more reliable, it brings a relative high cost of the Instrumentation and Control (I&C) system and a fairly complex layout of the plant. There are tens of thousands of devices in a nuclear power unit. Only a few important devices related with nuclear safety, such as the main pump, are under the comprehensive monitoring by DCS [6]. More than half of devices are out of the monitoring by DCS, due to the limited number of DCS input/output (I/O) channels. The I/O channel number is constrained not only by the cost of DCS modules, but also by the confined space occupied by DCS cabinets. Moreover, a large number of cables connecting devices over the plant would require more cable trays which may lead to space collision problems for design and construction.

Applying the wireless sensor network (WSN) in monitoring part of equipment and processes in NPPs can effectively alleviate the above problems. The wireless sensors are able to send monitoring information back to the control center via wireless channels. Therefore, the number of wired I/O channels of DCS could be reduced under the same monitoring requirements. The replacement of some wired cables with wireless connection will create the benefits of fewer cables and trays, less space for I&C equipment, and the shorter design and construction period.

Wireless sensor networks can not only lower the cost of regular monitoring, but also enable the capability to achieve higher levels of monitoring. With wireless sensor networks, massive and heterogeneous data about equipment and processes can be obtained. These rich monitoring data offer brand new possibilities for the fault detection and diagnosis, the advanced operation and maintenance, the physical and cyber security defense.

To achieve such high levels of intelligent monitoring, three questions need to be answered. Q1: what kinds of process variables and equipment parameters need to be included into monitoring? Q2: how the massive and heterogeneous data can be transmitted via the WSN? Q3: how do we deal with the large scale of data to realize intelligent monitoring?

The rest of this paper is organized as follows. What data to monitor in Q1 depends on intelligent monitoring schemes for various scenarios. Three monitoring schemes for different applications are proposed in Sect. 2. How to transmit data in Q2 is specified in wireless protocols. In Sect. 3, wireless protocols employed by the WSN are compared and other important issues are discussed. We also demonstrate specific data processing algorithms related with Q3 in Sect. 4. Conclusions are made in Sect. 5.

2 Intelligent Monitoring Schemes for Various Scenarios

Intelligent monitoring with rich information about processes and equipment will make a great benefit for safety and economy efficiency of NPPs. Currently, the amount and types of monitoring information adopted by NPPs are for achieving the control performance in normal operation or for actuating protections in abnormal occurrences. However, it has not given full considerations to aspects of cyber security, fault diagnosis, and smart maintenance. To realize such high level of intelligent monitoring, massive and heterogeneous data are required [10]. Multiple types of data out of the scope of DCS monitoring can be included into intelligent monitoring with the WSN, such as radio-frequency identification (RFID) tags data, video camera data, and vibration sensor data. Some examples will be elaborated below to show how the rich information could contribute to cyber-attack detection, fault diagnosis and recovering, and nuclear security.

To detect cyber-attacks on electrical motors in the Stuxnet scenario, heterogeneous observations besides the rotational speed (rpm) of a motor can be supplemented with the WSN. A motor under cyber-attacks which does not rotate smoothly will make unusual noises and eventually cause abnormal vibration of the motor case. Thus a variety of wireless sensors could be positioned around the motor to provide diverse information. An infrared thermometer placed above the motor can observe its temperature increase. A vibration sensor laid on the surface of the motor case can measure its vibration. A microphone deployed beside the motor can hear its noise. If there is a cyber-attack aiming to destroy the motor by disturbing its rotation, not only the rotational speed would go wrong, but anomalies will occurs in signals of the infrared thermometer, the vibration sensor, and the microphone as well. The diverse information about the motor could help to build anomaly detection against cyber-attacks.

Another example shows fault diagnosis and recovering with wireless monitoring data. The feed water flow in NPPs measured by the flowmeter could be send back to the process control system via the wired cables as well as the wireless channel. The wireless data serves as a diverse backup for the wired data. The backup data can verify the wired data to detect malfunctions of cables or control modules. If the fault is identified, the wireless data can substitute for the faulty data to maintain the continuous monitoring.

The third example demonstrates how the RFID technology and wireless networks can strengthen the security of nuclear material. The RFID tag is composed of memory, modulator, and wireless antenna. They can be attached on any nuclear material containers or packages that require rigorous monitoring. The information about nuclear material can be recorded and updated on the RFID tags. Through RFID reader, the information on tags can be read in a contact-less way. Meanwhile, the positions of nuclear material can be localized and tracked by RFID tags and readers pair. The up-to-data information retrieval and the localization by RFID can reduce risk of nuclear material loss.

3 Wireless Issues for the WSN in NPPs

To employ wireless technologies to transmit the monitoring data in NPPs, three major concerns need to be comprehensively considered. They are the electromagnetic compatibility of wireless devices, the security of wireless networks, and protocols selection for the requirements of wireless applications.

For the first concern, wireless devices must be electromagnetically compatible (EMC) with the existing I&C systems. The power level of wireless sensors is usually below 20 mW. In the EMC standard IEC 61000-4-3 (GB/T 17626.3), note 4 in Sect. 5.2 states: "Other systems operating in this frequency range, e.g. radio LANs operating at 2.4 GHz or higher frequencies, are generally very low power (typically lower than 100 mW), so they are much less likely to present significant problems."

For the second one, wireless transmission in opening space makes the network platform vulnerable to external malicious attacks. Thus wireless platform must be based on a private network. With the advanced technologies of encryption, authorization, channel fault-tolerance and security certification, the wireless network can be guaranteed to be reliable and secure.

For the third one, more and more applications have been or are going to be built on the wireless platform of NPPs, such as voice communication, personnel positioning, and radiation dose monitoring. The requirements of future wireless applications need to be taken into account in advance. The characteristics of the wireless platform in NPPs should be as follows.

1. High bandwidth;
2. Lower RF power, while covering a wide range;
3. Scalable and extensible with the capacity of sharing components;
4. Widespread use of communication protocols supported by industry.

Wireless technologies which have been employed in NPPs in China include: Wi-Fi, McWill, PHS and TDD (Table 1). Currently these technologies are mainly used for voice communication. The great potential of wireless data communication has not yet to be fulfilled. Only McWill and Wi-Fi technologies have the potential for the data communication. They have the properties of high bandwidth, high security, and terminal roaming handoff. However McWill causes higher RF power, thus only allows one-way communication for mobile receiver in nuclear island [7]. Wi-Fi technology with 2.4 GHz band meets all the above requirements of wireless applications in NPPs. The high bandwidth of communication, the openness of protocols and the scalability of the systems make the Wi-Fi suitable for a variety of applications, such as wireless monitoring, emergency communications and mobile operation. .

4 Data Processing Algorithms for Intelligent Monitoring

To fuse a large scale of data collected by the WSN to realize intelligent monitoring, specific data processing algorithms are required. Two statistical algorithms for fault detection and fault classification are demonstrated below.

Table 1 Comparison of wireless technologies in NPPs

Tech Spec.	McWill	WiFi	PHS	TDD
Frequency band	1875–1805 MHz Need licenses	2.4 GHz Free band	1900–1920 MHz Need licenses	800 MHz Need licenses
Bandwidth	15.36 Mbps	300 Mbps	64/128 Kbps	64 Kbps
Transmitting power	Base station 30 W (outdoor) Base station 10 mW (indoor) Cell phone 2 W (outdoor) Cell phone 10 mW (indoor)	Base station 0.1 W (indoor) Base station 0.5 W (outdoor) Cell phone 0.1 W	Base station 10 mW Cell phone 10 mW	Base station 0.6–40 W Cell phone 1.8 W
Coverage (outdoor) (km)	4	0.5	0.5	5
Mobility (km/h)	120	50	40	100
Advancement	Better	Great	Normal	Good
Security	Encryption, group function, authorization, channel fault-tolerance and security certification	Encryption, group function, authorization, channel fault-tolerance and security certification	AI encryption	Authorization, P2P encryption
Industrial cluster	Scarcely	More than 300 alliance member	Weed out	TETRA/PDT/GT800/GoTa/iDEN
Functionality	Dispatcher voice/note/data/real-time video	Dispatcher voice/note/data/HD real-time video/location	Voice/note	Voice/note

4.1 KPCA Algorithms for Fault Detection

Kernel principal component analysis (KPCA) performs a nonlinear mapping from the high dimensional input space to the lower dimensional feature space through kernel function. It greatly simplifies the calculation through converting inner product computation into kernel function calculation. In KPCA, $\mathbf{x} \in R^M$ are projected onto feature space $\Phi : R^M \to F$ through a nonlinear mapping function. The covariance can be expressed as,

$$\mathbf{C} = \frac{1}{N}\sum_{n=1}^{N} \Phi(\mathbf{x}_n)\Phi(\mathbf{x}_n)^T \quad (1.1)$$

where $\Phi(\mathbf{x}_n)$ is the n th sample in the feature space with zero mean and unit variance. N is the total number of the samples. Because $\Phi(\cdot)$ is usually hard to obtain, kernel matrix $\mathbf{K}$ is used,

$$k_{ij}=k(\mathbf{x}_i, \mathbf{x}_j) = \langle \Phi(\mathbf{x}_i), \Phi(\mathbf{x}_j) \rangle \quad (1.2)$$

Here, the radial basis function is used as the kernel function.

By calculating $\lambda\alpha = \mathbf{K}\alpha$, it yields the orthonormal eigenvectors $\boldsymbol{\alpha}_1, \boldsymbol{\alpha}_2, \ldots, \boldsymbol{\alpha}_N$, and the corresponding eigenvalues $\lambda_1 \geq \lambda_2 \geq \cdots \geq \lambda_N$. The dimensionality can be reduced by retaining only the first R eigenvectors. The score vector t_k can be obtained as,

$$t_k = \langle \mathbf{v}_k, \Phi(\mathbf{x}) \rangle = \sum_{n=1}^{N} \alpha_n^k k(\mathbf{x}_n, \mathbf{x}) \quad (1.3)$$

where $\mathbf{v} = \sum_{n=1}^{N} \alpha_n \Phi(\mathbf{x}_n)$.

To apply KPCA algorithm to condition monitoring and fault diagnosis of nuclear power plants, monitoring statistical control charts, Hotelling T^2 and SPE, need to be constructed to reflect the operating condition.

Hotelling T^2 statistic is the standard sum of squares of principal component vectors.

$$T^2 = [t_1, t_2, \ldots, t_p]\Lambda^{-1}[t_1, t_2, \ldots, t_p]^{\mathrm{T}} \quad (1.4)$$

where $\Lambda = diag[\lambda_1, \lambda_2, \ldots, \lambda_p]$. The Hotelling T^2 statistic confidence limit can be estimated by F-distribution,

$$T^2_{\mathrm{lim}} = \frac{R(N+1)(M-1)}{N^2 - NR} F_{\alpha,R,M-R} \quad (1.5)$$

where $F_{\alpha,R,N-R}$ is the upper $100\alpha\%$ critical point of the F-distribution with R and $M-R$ degrees of freedom, and α is the significance level.

SPE statistic is the error between the change trend of each sample and the statistical model. SPE statistic in the feature space is defined as:

$$SPE = \left\|\Phi(\mathbf{x}) - \hat{\Phi}_R(\mathbf{x})\right\|^2 = \sum_i^N t_i^2 - \sum_i^R t_i^2 \tag{1.6}$$

where $\hat{\Phi}_R(\mathbf{x}) = \sum_i^R t_i \mathbf{v}_i$. The SPE statistic confidence limit can be estimated as the following equation,

$$SPE_{\lim} = g\chi^2_{h,\alpha} \tag{1.7}$$

where $g = \frac{v}{2m}$, $h = \frac{2m^2}{v}$. m and v are the estimated mean and variance of SPE respectively.

Faults can be detected by judging whether T^2 and SPE statistics exceed the respective confidence limit or not. In recent years, KPCA has shown its capability of nonlinear process monitoring. KPCA was used to detect faults in two example systems [3, 8]. It was found that KPCA could effectively capture the nonlinear relationship between the process variables. The problem of fault detection was addressed in mechanical systems using a KPCA-based method [9]. In view of the nonlinear characteristics in practical process of NPPs, the KPCA technique provides a method for fault detection in NPPs, which is a topic that has not been explored in the literature. The architecture of the KPCA-based fault detection algorithm is shown in Fig. 1.

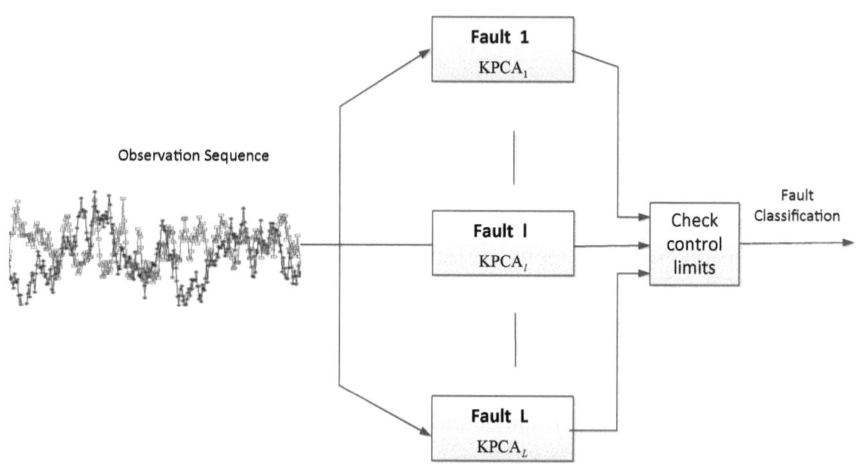

Fig. 1 Architecture of the KPCA-based fault detection algorithm

4.2 SVM Algorithms for Fault Classification

Support vector machine (SVM) is adopted to identify the faults. Define the training set $\mathbf{x}_i \in \mathbf{R}^n$ and $y_i \in \{+1, -1\}$ their class values. To construct the optimal hyperplane $H : \mathbf{w} \cdot \mathbf{\Phi}(\mathbf{x}) + b = 0$ that can separate the two-class samples, a non-negative relaxation variable $\xi_i \geq 0$ is introduced. To construct the optimal classification plane is to solve the following optimization problem:

$$\begin{aligned} \min \quad & \frac{1}{2}\mathbf{w}^T\mathbf{w} + C\sum_{i=1}^{n}\xi_i \\ s.t. \quad & y_i[\mathbf{w} \cdot \mathbf{\Phi}(\mathbf{x}_i) + b] \geq 1 - \xi_i, \quad i = 1, 2, \ldots, n \end{aligned} \tag{1.8}$$

where C is the penalty parameter. By using Lagrange function, the above problem can be transformed into

$$L(\mathbf{w}, b, \xi, \alpha, \beta) = \sum_{i=1}^{n}\alpha_i - \frac{1}{2}\sum_{i=1}^{n}\sum_{j=1}^{n}\alpha_i\alpha_j y_i y_j[\mathbf{\Phi}(\mathbf{x}_i) \cdot \mathbf{\Phi}(\mathbf{x}_j)] \tag{1.9}$$

where α_i is Lagrange multiplier. By replacing $[\mathbf{\Phi}(\mathbf{x}_i) \cdot \mathbf{\Phi}(\mathbf{x}_j)]$ by $K(\mathbf{x}_i, \mathbf{x}_j)$, it can be transformed into the dual quadratic programming problem as follows:

$$\begin{aligned} \max_{\alpha} \quad & \sum_{i=1}^{n}\alpha_i - \frac{1}{2}\sum_{i=1}^{n}\sum_{j=1}^{n}\alpha_i\alpha_j y_i y_j K(\mathbf{x}_i, \mathbf{x}_j) \\ s.t. \quad & \begin{cases} \sum_{i=1}^{n} y_i\alpha_i = 0 \\ 0 \leq \alpha_i \leq C, i = 1, 2, \ldots, n \end{cases} \end{aligned} \tag{1.10}$$

Given the test samples $\mathbf{x}$, classification function of SVM classifier can be written as:

$$f(\mathbf{x}) = \text{sgn}[\mathbf{w} \cdot \mathbf{\Phi}(\mathbf{x}) + b] = \text{sgn}[\sum_{i=1}^{n}\alpha_i y_i K(\mathbf{x}_i, \mathbf{x}) + b] \tag{1.11}$$

where sgn is the sign function.

When constructing classifier m, the output of Class m samples is trained as 1 and the output of the rest samples is trained as −1. The test sample x is first put into classifier 1. If the output of classification function $f^1(x)$ is 1, x belongs to Class 1. Otherwise, x is put into classifier 2. If the output of $f^2(x)$ is 1, x belongs to Class 2. Otherwise, x is put into classifier 3. The rest can be done in the same manner. When x is put into classifier $k-1$, if the output of $f^{k-1}(x)$ is 1, x belongs to Class $k-1$. Otherwise, x belongs to Class k (Fig. 2). SVM has been widely employed in fault classification. Four fault types of power transformer were identified by the trained

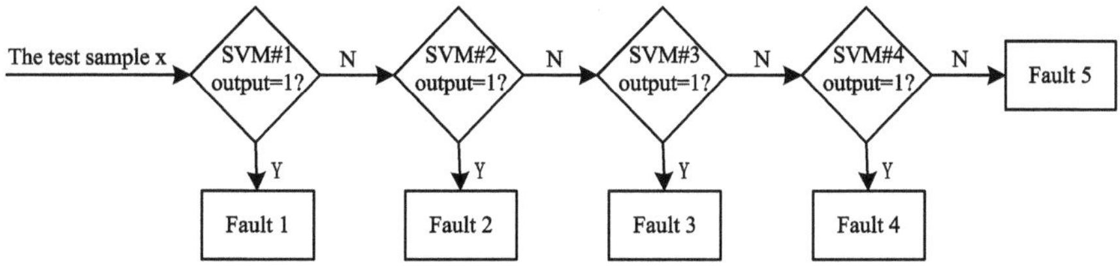

Fig. 2 Architecture of the SVM-based fault classification algorithm

multi-layer SVM classifier [4]. A hybrid two stage one-against-all SVM approach was proposed for fault diagnosis of defective rolling element bearings [5]. An improved SVM classifier was developed to fault diagnosis of actual analog circuits [2]. This paper is first concerned with fault detection and isolation problems for NPPs by combining KPCA and SVM. Thus, KPCA-SVM based algorithm is expected to realize the fault diagnosis of NPP with the WSN in the future work .

5 Conclusion

Although the wireless sensor network cannot substitute for the wired instruments and DCS at present, the WSN can serve as a diverse backup for non-safety I&C systems. Meanwhile, the WSN collecting massive and heterogeneous data is able to support a great many of intelligent monitoring applications, including the cyber security defense, the fault detection and diagnosis, and the security of nuclear material in NPPs. Three major concerns about applying wireless technology in NPPs are discussed. By comparison, the Wi-Fi technology is promising to be the underlying platform for wireless data communication in NPPs. Statistical data processing algorithms, KPCA-SVM based algorithms for anomaly detection and identification, are demonstrated. Specific algorithms for other applications need to be developed in the future work.

Acknowledgements This paper is jointly supported by the National S&T Major Project (Grant No. ZX06901) and National Natural Science Foundation of China (Grant No. 61502270).

References

1. Bi D, Zhang J (2010) Applications of wireless Technology in nuclear power plants and research on its key issue (in Chinese). Process Automation Instrumentation 31:47–53.
2. Cui J, Wang YR (2011) A novel approach of analog circuit fault diagnosis using support vector machines classifier. Measurement 44:281–289.
3. Elaissi I, Jaffel I, Taouali O et al (2013) Online prediction model based on the SVD–KPCA method. ISA Transactions 52:96–104.

4. Gan LV, Cheng HZ, Zhai HB et al (2005) Fault diagnosis of power transformer based on multi-layer SVM classifier. Electric Power Systems Research 7:1–7.
5. Gryllias KC, Antoniadis IA (2012) A Support Vector Machine approach based on physical model training for rolling element bearing fault detection in industrial environments. Engineering Applications of Artificial Intelligence 25:326–344.
6. Hashemian HM (2011) Wireless sensors for predictive maintenance of rotating equipment in research reactors. Annals of Nuclear Energy 38:665–680.
7. Jiang S, Chai B, Cao Y et al (2010) Study of the technical solution to the dedicated wireless communication system in nuclear power plant (in Chinese). Chinese Journal of Nuclear Science and Engineering 30:275–279.
8. Lee JM, Yoo CK, Choi SW (2004) Nonlinear process monitoring using kernel principal component analysis. Chemical Engineering Science 59:223–234.
9. Nguyen VH, Golinval JC (2010) Fault detection based on kernel principal component analysis. Engineering Structures 32(11):3683–3691.
10. Qin SJ (2012) Survey on data-driven industrial process monitoring and diagnosis. Annual Reviews in Control 36:220–234.

Performance Comparison of Tri-level Switching AMB Power Amplifiers Based on Different Current Sensors and Control Methods

Kai Zhang, Yang Xu and Jin-Ping Dong

Abstract Active magnetic bearings are advanced bearing components for High Temperature Gas-cooled Reactors. As an important part of active magnetic bearings, switching power amplifiers are designed to accurately convert a command signal to a real bearing coil current. Current ripples of the power amplifiers can be reduced effectively based on different tri-level control methods. Their performances are different. Even when the same control method is used, the performances of amplifiers using electrical circuits with different hardware components are different. To reduce current ripples of amplifiers, the performance of amplifier circuits with different current sensors and different control methods were studied. With the same control method, three kinds of current sensors were used in a power amplifier and their performances were compared. It was shown that hardware delay caused the performance difference. With the same current sensor, a tri-level Sampling-Hold method and a Pulse-Width Modulation (PWM) method were used to control the same amplifier driver. The accuracy of the tri-level PWM amplifier was better. The reason for the performance difference between the two methods was discussed. The experimental results were provided.

Keywords High temperature gas-cooled reactor · Pulse-width modulation · Power amplifier active magnetic bearing

1 Introduction

Because of their advantages of no wear, no oil, high speed and low maintenance cost, active magnetic bearings (AMB) have been widely used in many industry applications such as vacuum pumps, grinding machines, flywheels, turbo machines

---

K. Zhang (✉) · Y. Xu · J.-P. Dong
Department of Engineering Physics, Tsinghua University, Beijing 100084, China
e-mail: zhangkai@mail.tsinghua.edu.cn

Y. Xu (ed.), *Nuclear Power Plants: Innovative Technologies for Instrumentation and Control Systems*, Lecture Notes in Electrical Engineering 400,
DOI 10.1007/978-981-10-3361-2_21

and so on [1, 3]. They are also used in the helium fan for High Temperature Gas-cooled Reactors [4] as an irreplaceable advanced bearing technology.

As an important part of AMB, a power amplifier is used to convert a current command signal to a real current in a magnet coil. The current makes the magnet producing a magnetic force to control the movement of a rotor with ferromagnetic material. In an AMB system, power loss of its power amplifiers can't be ignored. Generally, a switching power amplifier is preferred over a linear power amplifier considering system power loss. But current ripples in a coil should be reduced effectively to obtain a high accuracy control current [5, 7, 8].

Traditional switching power amplifiers are controlled by two-level methods including the two-level Pulse-Width Modulation (PWM) method, the two-level Hysteresis method, the two-level Sampling-Hold method and the two-level Minimum Pulse-Width (MPW) method [5, 7, 8]. For the two-level amplifiers, the voltage on the driven coil is a DC link voltage (positive or negative) and the coil is in a charging state or a discharging state. The current ripples of a two-level amplifier are large when the DC link voltage is increased to achieve a high bandwidth for the amplifier. They cause low current accuracy and high power loss [2, 6, 9]. To reduce the ripples, tri-level amplifiers are developed in AMB application. Compared with the two-level amplifiers, an idle state is added in a tri-level power amplifier to smooth the coil current.

The work is focused on the influence of hardware components and control methods of power amplifiers. The influence of the hardware components is mainly in the dynamic response speed. For the electrical circuit of an amplifier, the most sensitive component for the response speed is the current sensor used. When different current sensors with the different respond speed were used, the corresponding ripple performance was studied. It was shown that the hardware delay caused the performance difference. At the same time, the control methods used also influence the ripple performance. With the same current sensor, a tri-level Sampling-Hold method and a Pulse-Width Modulation (PWM) method were used to control a current amplifier. Their performances were compared. The reason for the performance difference between the two methods was discussed. The experimental results showed that the tri-level PWM amplifier could achieve better current accuracy.

2 Working Principle of a Tri-level Power Amplifier

Compared with a two-level amplifier, a new state called idle-state is added in a tri-level amplifier. In an idle-state, the voltage on the coil driven is nearly zero; the resistor in the coil current loop is low and the current flows with very low power loss. The topology for the output stage can be a half bridge or a full bridge. They can both used to achieve a tri-level current power amplifier. The full bridge can be even used to drive a bi-directional current [2].

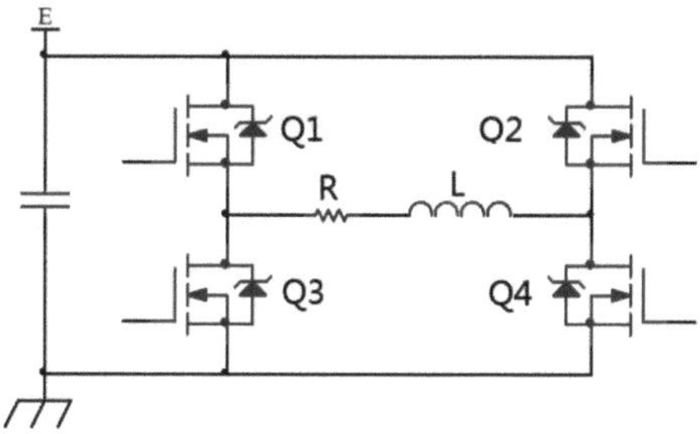

Fig. 1 Structure of the full bridge topology circuit

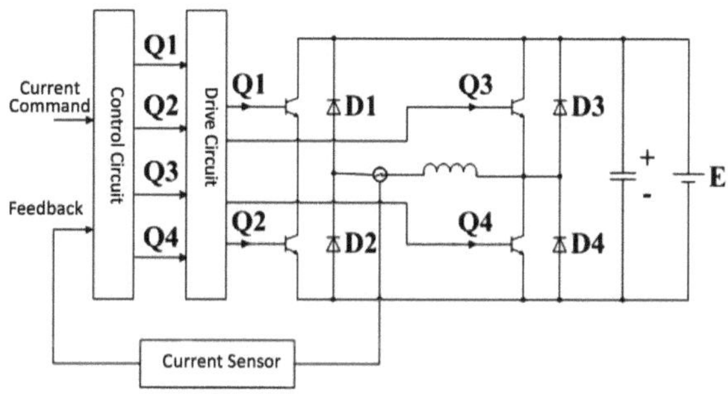

Fig. 2 Structure of the amplifier circuit

In Fig. 1, a simplified structure of the full bridge topology circuit is shown. The bridge contains four power transistors: Q1, Q2, Q3 and Q4. A magnet coil is in the middle of the bridge.

3 Realization of a Tri-level Power Amplifier

There are different kinds of tri-level power amplifiers developed before, such as tri-level Sampling-Hold amplifiers, tri-level Hysteresis amplifiers and tri-level PWM amplifiers. These amplifiers have a similar structure as Fig. 2. Their differences are in the control circuit (in Fig. 2) which is designed based on a special control method and used to produce switch signals for the power transistors of the bridge.

The work about the control methods is focused on the two popular tri-level amplifiers: the tri-level PWM amplifier and the tri-level Sampling-Hold amplifier. They are the most used tri-level switching power amplifiers in magnetic bearing systems.

3.1 Tri-level Sampling-Hold Amplifier

A tri-level Sampling-Hold switching amplifier is based on a traditional two-level Sampling-Hold switching amplifier, and its switch control signals are produced by specific designed timing logic modules or a truth table.

There is a clock signal for the control circuit of the amplifier. It determines the working frequency of the power transistors. At every rising edge, the error of the command signal and the measured current signal is sampled. Based on the sign of the error, the control circuit produces four switching signals for the four bridge power transistors separately and makes the coil enter a charging state or a discharging state. During this clock period, the error is monitored continually. If the sign of the error changed (from positive to minus or from minus to positive), the coil enters an idle state until the next rising edge of the clock. Such a control scheme insures that the power transistors won't switch more than once within a clock frequency and the switching frequency of the transistors can be reduced. Such an amplifier has the advantages of high stability, wide bandwidth, simple circuit and low current ripple [2].

By reducing the switch frequency, the electromagnetic interference from high frequency switch signals can be reduced. But the control scheme still has its disadvantages. When the change of the error sign just comes from a noise signal, the amplifier will enter an idle state because of a misjudgment until the next rising edge. It reduces the response speed of the amplifier and increases the current ripples.

3.2 *Tri-level PWM Amplifier*

A PWM amplifier uses a pair of bipolar triangular waves to compare with the current error signal and gets the corresponding switch control commands. The time curve of the error signal is divided into three regions by the triangular waves as shown in Fig. 3. Region 1: the error signal is larger than the positive triangular wave and the coil is in a charging state; region 2: the error signal is smaller than the minus triangular wave and the coil is in a charging state; region 3: the error signal is smaller than the positive triangular wave, larger than the minus triangular wave and the coil is in a idle state. The control circuit is sensitive when the error is near the cross point of the two triangular waves (e.g.: position α in Fig. 3). The amplifier usually charges or discharges here. When the error comes to the middle position of

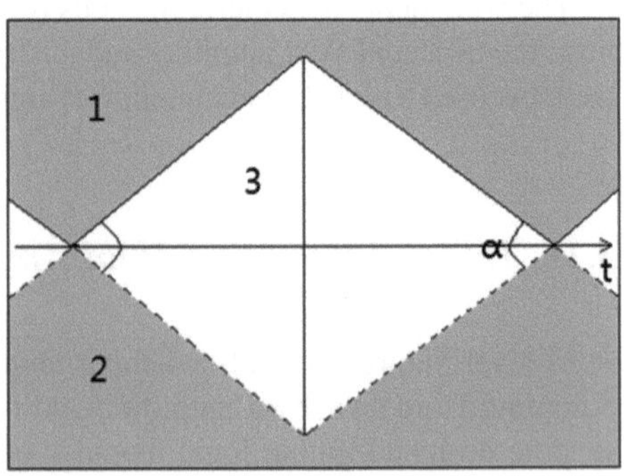

Fig. 3 Region division for the time curve of the error signal

region 3, error tolerance can be relatively larger and the amplifier is usually in an idle state. The tri-level PWM control scheme has the advantages of high stability, low current ripple and controllable switching frequency [2, 9].

4 Influence of Current Sensors

Many researchers have discussed the performance difference between tri-level amplifiers and two-level amplifiers [2, 5]. For one special tri-level amplifier of AMBs, it is seldom discussed what happens when time delay of hardware used in the amplifier changes.

The most sensitive component for the response speed is the current sensor. For one tri-level Sampling-Hold amplifier working with a clock of 40 kHz, different current sensors were used to feedback the coil current signal. These sensors had different time delay and they were used to study the influence of the response speed. The test coil used for the amplifier has a register of 3.4 Ω and an inductance of 3.6 mH. The DC link voltage used is E = 28 V.

There were three kinds of current sensors being used in the amplifier. They were a TamuraL18P hall current sensor (Tamura Sensor), a LEM LA28-NP hall current sensor (LEM Sensor) and an AD 8211 voltage current sensor (Voltage Sensor). The Tamura Sensor and the LEM Sensor were both hall current sensors. They measured the current flowing in a coil through hall-effect. The Voltage Sensor obtained the current value by measuring the voltage on a sampling resister connected with the coil in serious.

The response speeds of the three sensors were different. The Voltage Sensor was the fastest and the LEM Sensor was the slowest. When the command signal for the amplifier was +1 V (1A command) DC signal and the amplifier worked with the three sensors respectively, the corresponding coil current ripple waves were recorded by an oscilloscope as shown in Fig. 4. It was seen that the amplifier achieved the smallest current ripple when the Voltage Sensor was used. If not considering transient switch noise in the current signal, a current accuracy of 40 mA peak-to-peak value was obtained. As a comparison, the current accuracy was about 60 mA for the LEM Sensor and 40–60 mA for the Tamura Sensor. When the Tamura Sensor was used, the ripple waves had two looks as c1 and c2 in Fig. 4 respectively.

If the amplifier would enter a charging state, the corresponding transistors should be opened. When the three sensors were used to provide the feedback signal respectively, the relationship between the error signal and the switch-on signal in the time domain was shown in Fig. 5. In the figure, the curve 1 was the output signal of the corresponding current sensor and the curve 2 was the sign signal of the current error. When the sign signal was high, it meant that the current in the coil was higher than the command signal. From the curve 1, it was easy to identify the moment that the transistor opened. The time between the transistor opening and the sign signal changing was called the inverting time and labeled as T in Fig. 5.

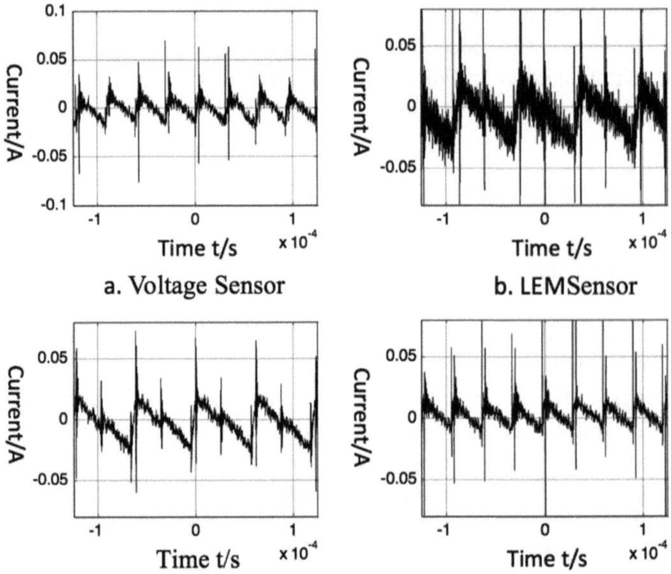

Fig. 4 Current ripple waves of tri-level sampling-hold amplifier with different current sensors

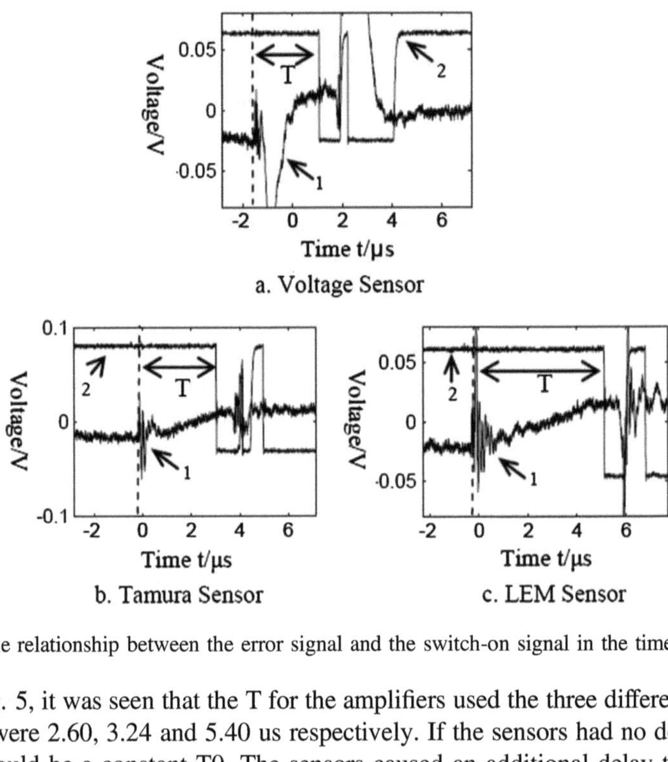

Fig. 5 The relationship between the error signal and the switch-on signal in the time domain

In Fig. 5, it was seen that the T for the amplifiers used the three different current sensors were 2.60, 3.24 and 5.40 us respectively. If the sensors had no delay time, the T should be a constant T0. The sensors caused an additional delay time Tlag.

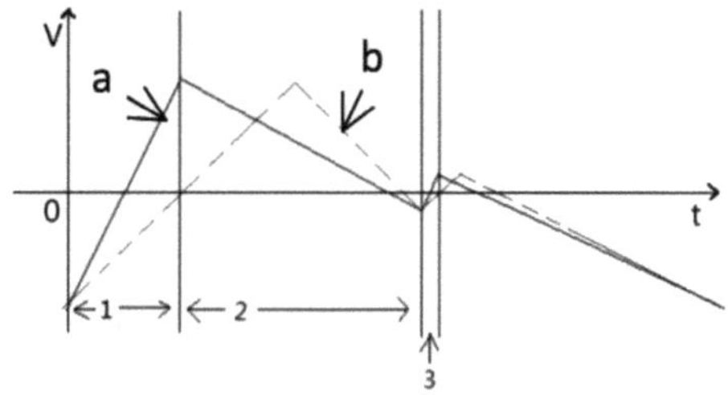

Fig. 6 Typical current waves for a tri-level amplifier with a current sensor

The total T was the sum of T0 and Tlag. Obviously, the Tlag for the three sensors were different. Tlag for the LEM Sensor was the largest. In its first charging state, the transistor could not close in time because of Tlag. When the next rising edge of the clock came, the current was only a bit lower than its command signal. It caused a very short adjacent charging time and the control circuit let the current move to an obvious minus current peak in the following idle state. The current curve looked as working at a half clock frequency as shown in Fig. 5b. The delay of the Voltage Sensor was the smallest; its positive current peak was smaller than the LEM Sensor. When the next rising edge came, the current had been obviously below the command signal and an obvious charging state followed after that. It produced the lowest current ripple level and worked at the clock frequency. The time delay of the Tamura Sensor was larger than the Voltage Sensor and smaller than the LEM Sensors. So its current ripple level was within the level of the other two sensors. The similar phenomenon could be seen in the working frequency as shown in Fig. 5.

To further illustrate the influence of the delay on current curves, a typical curve was divided into three regions in Fig. 6. In Fig. 6, the curve “a” was a real current error curve and the curve “b” was the current error measurement curve of a current sensor. In the region 1, the amplifier was in a charging state and the current curve rose quickly. Because the initial error was relatively large, the charging time was relatively long. The delay time between the zero-crossing time of the curve “a” and the curve “b” was obvious. It caused an overshoot of the current curve “a”. In the region 2, the amplifier was in an idle state and the current curve changed slowly. So the sensor could more easily follow the current signal. Because the overshoot in the region 1 was relatively large and the current changed slowly in the region 2, at the end of the idle state in the region 2, the error value of the current was small. After the idle state, the amplifier began a new charging state as the region 3. In the region 3, because of the small error value, the curve “a” reached zero in a short time; the curve “b” reached zero too soon after that. When the curve “b” reached zero, the control circuit let the charging state end and the amplifier began a new idle state. The idle state would not end until the coming of the next clock rising edge. Then a new large initial error would come and the amplifier would start a new circle. If the region 3 was ignored and the two closing idle states were combined, the working frequency of the amplifier looked just like half of the clock frequency.

5 Influence of Control Methods

The control methods used also influence the ripple performance. The tri-level Sampling-Hold method and PWM method were compared when they were used to control the same amplifier with the Tamura Sensor. They worked based on the same bridge stage and the same current sensor.

When the command signal was a +1 V (1A command) DC signal, the experimental current waves were shown as Fig. 7. If the high frequency switching noise was ignored, it was shown that the tri-level PWM method obtained a smaller current ripple and higher current accuracy compared with the tri-level Sampling-Hold method.

Based on the different working principle compared with the Sampling-Hold method, the PWM method could monitor the current error signal in a whole clock period and it could more effectively reduce noise disturbance. Moreover, there was a good deviation correction effect in the PWM method when it controlled the transistors based on the comparison results between the command signal and the triangular waves. Without loss of generality, as shown in Fig. 8, the error signal changed from zero at the beginning; the triangular wave changed with a constant velocity of V1; it rose in the region "a" and dropped in the region "b"; the error signal changed with a velocity of V2(t). V1 could be used as a reference to limit V2

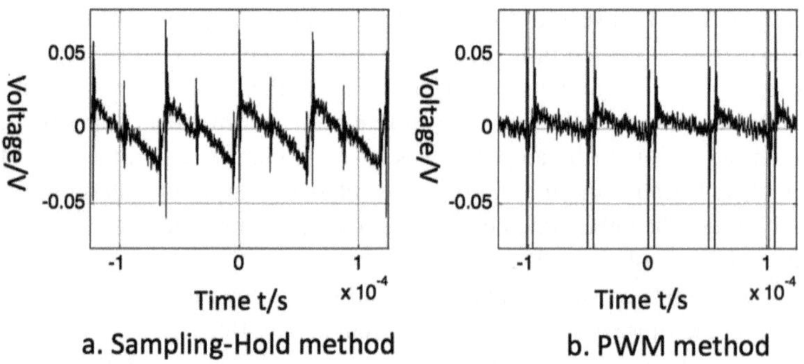

Fig. 7 Current waves of the two control methods

Fig. 8 Deviation correction effect in the PWM method

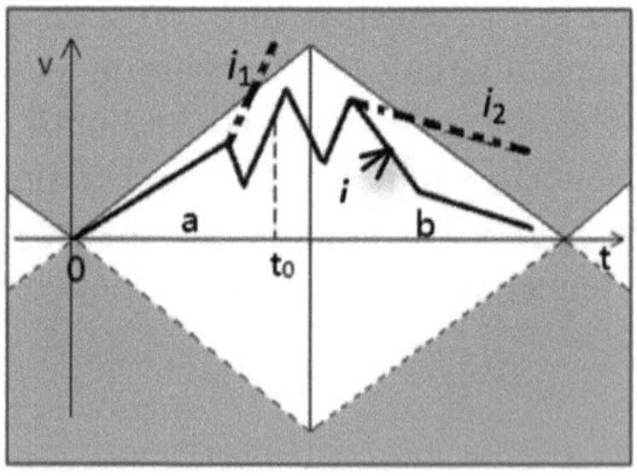

and the integration of V2 in the time domain (corresponding to the average velocity). In the region "a", if the formula (1) was satisfied, the amplifier stayed in an idle state. On the contrary, if the current error increased rapidly as i1 in the figure, the control circuit would change to a recharging state.

$$\left|\int_0^{to} V_2(t)dt\right| < |V_1 to| \tag{1}$$

In the region "b", the triangular wave dropped with a constant velocity of −V1. If the average velocity magnitude of the error signal was less than V1 as i2 shown in Fig. 8, the control circuit would change from the idle state to a recharging state. With the deviation correction effect, the amplifier allowed the current having a small fluctuation to decrease the corresponding switching frequency. At the same time, a large fluctuation could be restrained in time to assure a high quality current wave. If the error signal was compared with the minus triangular wave, similar things happened.

When the tri-level Sampling-Hold amplifier stayed in a charging or discharging state, it closed the opened transistors and entered an idle state once the coil current reached the command value. If there was a non-ignorable time delay in the circuit,

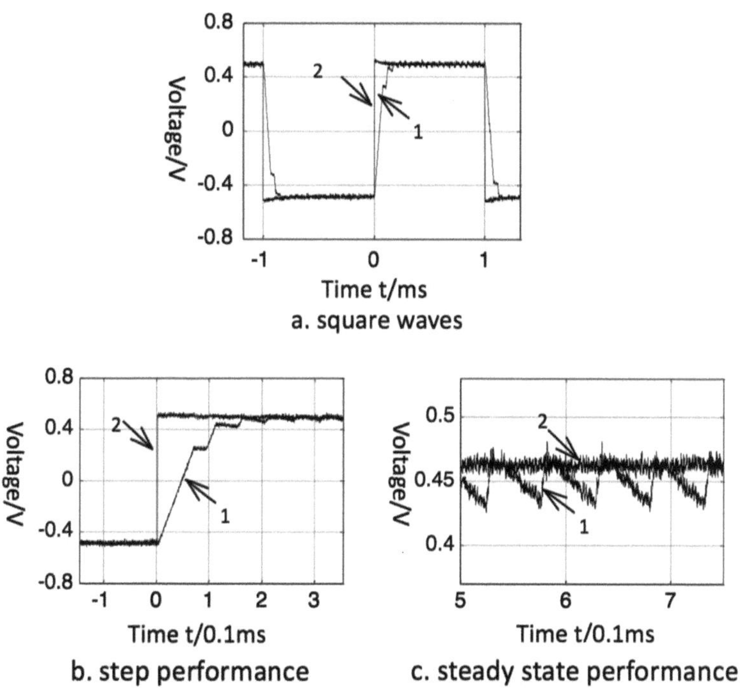

Fig. 9 Current waves of the PWM amplifier

the delay would cause an additional current overshoot. For the tri-level PWM amplifier, as shown in Fig. 8, the control circuit could monitor the error signal during the whole clock frequency; the deviation correction effect could correct the misjudgment from noise signal; the influence of the time delay and some other unexpected circuit noise could be reduced and the current quality could be improved.

The PWM method was tested in a test amplifier. When the reference signal was a square signal with a frequency of 500 Hz and peak-to-peak amplitude of 1 V, the reference signal and the current signal were measured and shown in Fig. 9. In the figure, the curve 1 was the current signal and curve 2 was the reference signal. In Fig. 9b, it was seen that the amplifier's good performance in tracking a signal step. After 3 charging periods, the current reached the command signal and the time used was about 162 us. In Fig. 9c, the waves were recorded when the square signal stayed at a constant voltage. It showed that the current ripples of the amplifier were small.

6 Experimental Comparison

To compare the performance of the Sampling-Hold method and the PWM method, two control circuits based on the two methods respectively were designed and manufactured. They were used to control the same power driver module. The power driver module consisted of an Intelligent Power Module (IPM) output stage, a current sensor and other necessary interface circuits. When the two control methods were used to construct the corresponding amplifiers, the amplifiers had the same hardware except the different control circuits. It was a good way to directly compare the performance of the two control methods.

In experiments, the amplifiers were used to track a constant reference signal for 1 A. The current ripple comparison was shown in Fig. 10. The current ripples of the Sampling-Hold method were shown in Fig. 10a and the corresponding

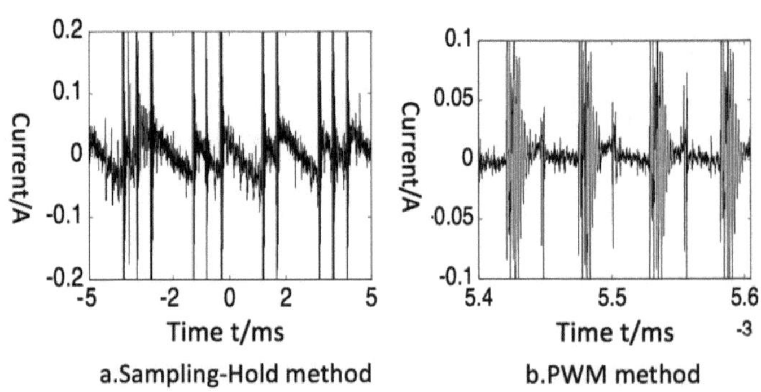

Fig. 10 Current ripple comparison

peak-to-peak value was about 200 mA. The current ripples of the PWM method were shown in Fig. 10b and the corresponding peak-to-peak value was about 40 mA. It was shown that the current ripples of the PMM method were much lower than that of the Sampling-Hold method.

7 Conclusions

For the tri-level switching power amplifiers of AMBs, the amplifier performance was influenced by its hardware speed and the control methods used. The influence of the hardware components was checked by three kinds of current sensors. They were used for the same amplifier and the current ripples obtained respectively were obviously different. With the same current sensor, the tri-level Sampling-Hold method and the Pulse-Width Modulation (PWM) method were used to control the same amplifier circuit. The performance comparison showed that the accuracy of the tri-level PWM method was much better than that of the tri-level Sampling-Hold method. The reason for the performance difference was discussed. The deviation correction effect of the PMW method could assure a high quality current wave. The experimental results also proved it.

References

1. Gerhard Schweitzer, Eric H. Maslen, etc. Magnetic Bearings: Theory, Design, and Application to Rotating Machinery [M]. Dordrecht, New York, Springer, 2009
2. LI Bin, Deng Zhiquan, Yan Yangguang. Current Mode Switching for Tri-state Amplifiers in Magnetic Bearing Control System Supporting High Speed Motors [J], Power Electronics, 2003, 37(4): 52–55
3. Schweitzer G. Active Magnetic Bearings: Basic, Properties, and Applications of Active Magnetic Bearings [M]. ETH Zurich, Switzerland: Verlag der Fachvereine Hochschulever-lag AG an der ETH Zurich, 1994
4. Shi Lei, Zhao Lei, Yang Guojun, etc. Design and Experiments of the Active Magnetic Bearing System for the HTR-10, 2nd International Topical Meeting on HIGH TEMPERATURE REACTOR TECHNOLOGY, Beijing, China, September 22–24, 2004
5. Zhang Guangming, Chen Chun, Mei Lei, etc. The Future and Development of Power Amplifier for Magnetic Bearings [J], Small & Special Electrical Machines, 2012, 40(3): 73–76
6. Zhang Jing, et al. Synchronous Three-level PWM Power Amplifier for Active Magnetic Bearings [C]. Proceedings of the 5th International Symposium on Magnetic Bearing, Kanazawa, Japan, 1996: 277–282
7. Zhang Liang Fang Jiancheng. Analysis of Current Ripple and Implementation of Pulse Width Modulation Switching Power Amplifiers for Active Magnetic Bearing [J], Transactions of China Electrotechnical Society, 2007, 22(3): 13–20
8. Zang Xiaomin, Wang Xiaolin, Qiu Zhijian, etc. Research on Current Mode Tri-state Modulation Technology in Switching Power Amplifier for Magnetic Bearings [J], Proceedings of the CSEE, 2004, 24(9): 167–172
9. Zeng Xueming, Xu Longxiang, Liu Zhengxun. Study of Three Level PWM Power Amplifier for AMB [J], Power Electronics, 2002, 36(3): 13–15

The Independence of Safety Digital I&C System in Nuclear Power Plant

Xiang Jia, Zhong-Qiu Wang, Yun-Bo Zhang and Yin-Hui Guo

Abstract Independence is one way to improve the reliability of digital instrumentation and control (I&C) system in nuclear power plant, and relevant regulation and standards give clear requirements about independence. Based on these regulation and standards, this paper briefly introduces the three means to achieve independence: electrical isolation, physical separation, communication isolation. Then it summarizes five aspects of the independence of digital I&C system. It is useful to the design and review of the digital I&C system.

Keywords Nuclear power plant · Digital I&C system · Independence

1 Introduction

For the design of safety system in nuclear power plant, in order to improve the system reliability, it usually adopts the redundancy and the diversity principle. Independence is an important way to achieve effective redundancy and diversity.

X. Jia
Ministry of Environmental Protection, Beijing, China
e-mail: jia.xiang@mep.gov.cn

Z.-Q. Wang · Y.-B. Zhang (✉) · Y.-H. Guo
I&C Department, Nuclear and Radiation Safety Center, Beijing, China
e-mail: zhangyunbo@chinansc.cn

Z.-Q. Wang
e-mail: wangzhongqiu@chinansc.cn

Y.-H. Guo
e-mail: Gyh86126@126.com

Y. Xu (ed.), *Nuclear Power Plants: Innovative Technologies for Instrumentation and Control Systems*, Lecture Notes in Electrical Engineering 400,
DOI 10.1007/978-981-10-3361-2_22

2 The Independence Means

HAF 102-2004 "nuclear power plant design and safety requirements" describes the redundancy and independence of protection system in nuclear power plant should be at least sufficient to ensure that [1]:

No single failure results in loss of protection function;
The removal from service of any component or channel does not result in loss of the necessary minimum redundancy, unless the acceptable reliability of operation of the protection system can be otherwise demonstrated.

HAF 102-2004 also points out that the reliability of system can be improved by maintaining the following features for independence in design:

1. Independence among redundant system components;
2. Independence between system components and the effects of postulated initiating events (PIEs) such that, for example, a PIE does not cause the failure or loss of a safety system or safety function that is necessary to mitigate the consequences of that event;
3. Appropriate independence between or among systems or components of different safety classes; and
4. Independence between items important to safety and those not important to safety.

GB/T13286 "The Independence Guidelines for Safety Electrical Equipment and Circuits in Nuclear Power Plant" provides the guidelines for the independence of safety electrical equipment and circuits [2].

Independence requirements can be reached by the following means.

2.1 The Electrical Isolation

The electrical isolation between the multiple parts of the system is used to reduce the possibility of adverse interactions. Typically design measures of electrical isolation include the use of isolation amplifier, control switch, current transformers, optical couplers, relays, circuit breakers and other equipment. GB/T13286 gives the acceptable isolation device (Table 1).

2.2 The Physical Separation

The physical separation is used to reduce the possibility of the loss of the multiple parts in the safety system caused by certain types of PIEs (such as fires, chemical explosions, plane crashes, flying objects, flooding, extreme temperature and

Table 1 Acceptable isolation device

Isolation device	
1	Amplifiers
2	Control switches
3	Current transformers
4	Fiber optic couplers
5	Photo-optical couplers
6	Relays
7	Transducers
8	Power packs
9	Circuit breakers

Table 2 Physical separation

	Physical separation includes
1	Separation by geometry (such as distance or orientation)
2	Separation by barriers
3	Separation by a combination of these

humidity, etc.) and failures at the same time. It usually adopts geometric partition (such as distance, direction, etc.), separated by a barrier or a combination of both methods to achieve physical separation. Physical separation usually is applied to redundant channels [3] (Table 2).

2.3 The Communication Isolation

The communication isolation can be used to prevent the spread of failures along the communications path in redundant channels or between the safety component and non-safety component, and prevent the loss of safety functions due to communication activity. GB/T13629 gives the means to meet the provisions of the communication independence, and makes recommendations for communication isolation methods. One is communication between computers in different safety channels. Another is communication between safety and non-safety computers [4].

3 The Independence of Digital I&C System

For digital I&C system, it usually considers the following independence.

3.1 The Independence in Reactor Trip (RT) System and Engineered Safety Features (ESF) System

Reactor protection system mainly includes reactor trip (RT) system and engineered safety features (ESF) system [5]. RT system and ESF system usually adopt redundant structure to meet the single failure criterion of safety system. Electrical isolation and physical separation methods in their redundant structure design are used to meet the independence requirements. Because of the data communication between multiple safety systems, RT system and ESF system also should meet the communications isolation requirement. The deterministic feature of software in safety system can ensure the independence of the multiple safety system.

When protection system (RT system and ESF system) share the analog signals with other systems, the electrical isolation is necessary to meet the independence requirement.

It also should be meet the independence between the protection system and the maintenance system.

3.2 The Independence Between the Diverse Actuation System and RT System

The diverse actuation system (DAS) should be diversity to the RT system. For independence, the application of analog equipment can overcome common cause failure, and it also can adopt the diverse digital equipment to protection system. When RT system share the analog signals with DAS system, the electrical isolation is necessary to meet the independence requirement [6].

Anticipated transients without trip (ATWT) mitigation function is one function of the DAS system, it should use different trigger variables or adopt diverse transmitter with RT system.

3.3 The Independence Between the Control System and Protection System

When protection system share the analog signals with control system, the electrical isolation is necessary to meet the independence requirement. At the same time, the communication isolation between them should be met. The design of independence should use isolation devices between components and use different cable and power circuit. For example, the communication between protection system and control

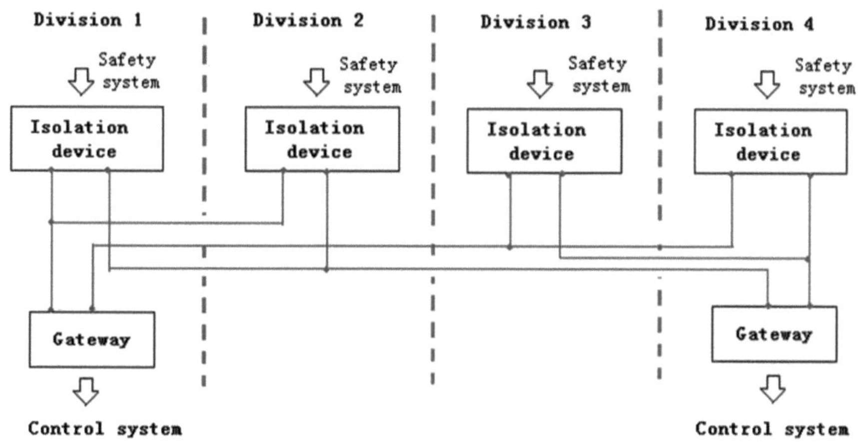

Fig. 1 Communicate between protection system and control system

system use isolation device to reach communication isolation, and the classification of this device should be safety. The use of isolation device can ensure the independence of the protection system and control system (Fig. 1).

3.4 The Independence of Manual Control and Display

Manual control and display usually adopt analog device in nuclear power plant, the signal which is shared with digital system should be electrical isolation to meet the independence requirement. In order to improve the availability of the plant, manual backup is usually provide to operator in the loss of digital device, so the plant can be maintain in stable station over several hours.

The minimum safety manual control and display should include accident and post-accident important information and the manual trigger of protection systems.

3.5 The Independence of Redundant Division of Safety System

Redundant division of safety system shall be independent. The independence between the redundant divisions can be met by the installation of digital I&C cabinets in different rooms. It can ensure the capability of accomplishing the safety function to a certain extent (Fig. 2).

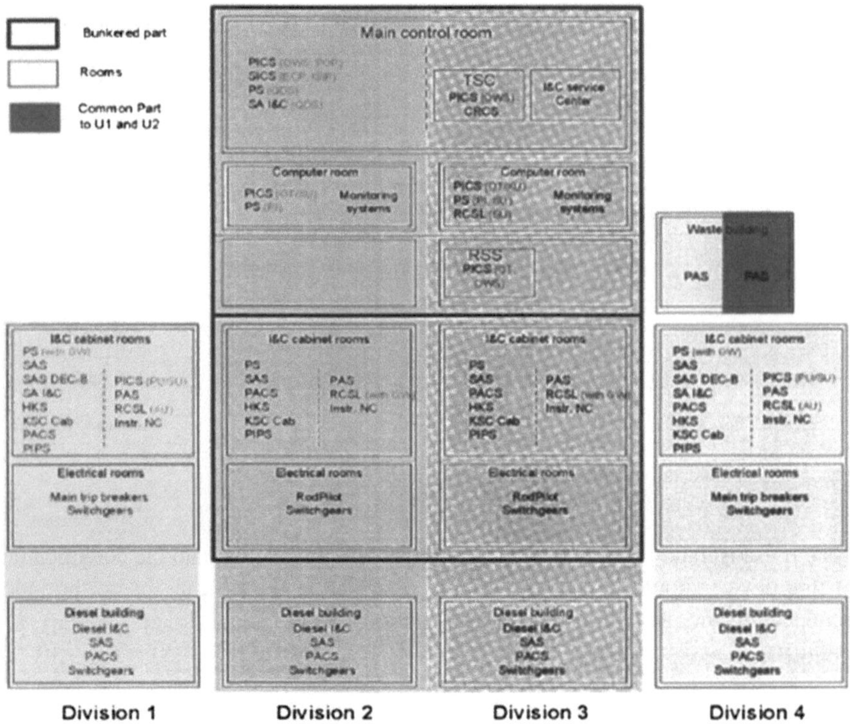

Fig. 2 The layout of digital I&C cabinets

4 Conclusion

Digital I&C system is the main development trend of nuclear power plant. Independence is an important requirement to the safety of digital I&C system. The main means of independence include electrical isolation, physical separation and communication isolation. In the design, it should be consider the independence of the independence in RT system and ESF system, the independence between the DAS and RT system, the independence between control system and protection system, the independence of manual control and display, the independence of redundant division of safety system.

References

1. HAF102-2004, Safety Requirements of Nuclear Power Plant Design.
2. GB/T13286-2008, the Independence Guidelines for Safety Electrical Equipment and Circuits in Nuclear Power Plant.

3. LU Chao (2012) Independence Design of Safety Class DCS System in Nuclear Power Plant, Nuclear Science and Engineering. Vol. 32 Suppl. 2: 232–237.
4. GB/T13629, Applicable criteria for digital computers in safety systems of nuclear power plants.
5. ZHENG Weizhi (2012) The Fault Analysis and Application of Reactor Protection System in Nuclear Power Plant. Nuclear Electronics & Detection Technology, Vol. 32. No. 3: 337–341.
6. ZHANG Yunbo (2014) Analysis of Diversity and Independence for ATWT Mitigation System in Nuclear Power Plant. Nuclear Power Engineering, Vol. 35. No. 6: 77–79.

Study on Application of Verification and Validation to Digital Control Room of Nuclear Power Plants

Yong-Bin Sun, Tao Bai and Li-Sheng Hu

Abstract Control rooms are the operational decision-making centres of nuclear power plants, by which the plant performance and safety of operation would be directly affected. It is estimated that almost 90% events and accidents happened in nuclear power plants are caused by operator failures. Therefore, how to effectively and efficiently maintain the plant's safety and performance by the modern digital control rooms is a critical topic due to their centralized and stacked display and control modes. From the viewpoint of verification and validation (V&V), the V&V process and methods are summarized and some V&V key issues are focused in this paper to ensure the quality of control rooms design.

Keywords Verification and validation (V&V) · Digital control room · Computerized graphic human system interface (HSI) · Computerized procedure · Computerized alarm system

1 Introduction

Nowadays, more and more nuclear power plants (NPPs) utilize digital instrumentation and control (I&C) systems to improve plant performance and maintain higher levels of safety, because of many advantages like self-checking, on-line diagnostics, improved accuracy and fault tolerance. As an integral part of the I&C systems, the control rooms with human-system interfaces (HSI) are also equipped by modern digital equipment, such as large screen displays, video display units (VDU) and so

Y.-B. Sun (✉) · L.-S. Hu
Department of Automation, Shanghai Jiao Tong University, Shanghai, China
e-mail: sunyongbin@cgnpc.com.cn

Y.-B. Sun
China Techenergy Co. Ltd., Beijing, China

T. Bai
State Key Laboratory of Nuclear Power Safety Monitoring Technology and Equipment, China Nuclear Power Design Co. Ltd., Shenzhen, China

Y. Xu (ed.), *Nuclear Power Plants: Innovative Technologies for Instrumentation and Control Systems*, Lecture Notes in Electrical Engineering 400,
DOI 10.1007/978-981-10-3361-2_23

on. Different from the traditional discrete analogy controls and displays, the modernized digital control rooms have many new characteristics, including integrated displays, VDU based information systems, soft controls, on-line computerized procedures, advanced alarm systems and so on, which should be considered in human-system interfaces (HSI) design.

The main control room in NPPs is the most important control centre, where the operators monitor and control the overall plant performance with safety and reliability. On one side, any subtle faults or weaknesses in HSI design of the main control room maybe cause the safety threat to the NPP. On the other side, it is shown by all of the three severe accidents, Three Mile Island (TMI) accident in the U.S. (1979), Chernobyl Accident in the former Soviet Union (1986) and Fukushima accident in Japan (2011) that the NPPs are thrown into the dangerous situations because of human factor failures. Therefore, verification and validation (V&V) is the most useful and effective way during the life cycle of the digital control room design. It could ensure that the HSIs, computerized graphic display and control layout, and procedures would meet the task and operators performance requirements, and be coincident with human cognitive and physiological characteristics as well. What is more, it would improve the operation reliability of the digital control room with its HSI.

In this paper, the V&V process and methods for design of digital control rooms are summarized according to NUREG 0700 [1], NUREG 0711 [2] and some engineering practices in Sect. 2. Some V&V key issues on digital control rooms are focused and discussed in Sect. 3 to ensure the quality of control rooms. Some application practices are given in Sect. 4. Finally, Sect. 5 concludes the paper.

2 V&V Process and Methods for Digital Control Room

According to NUREG 0700 [1], NUREG 0711 [2] and some engineering practices [3] such as the digital control room V&V for LING-AO nuclear power engineering, V&V process for design of the digital control rooms is summarized, shown as Fig. 1. It could be divided into three phases, including Preparation, Evaluation, Resolution and Regression.

In the first phase (Preparation), the V&V plan and methodologies should be prepared, where evaluation criteria, V&V methods and tools, V&V input files, V&V team and personnel, and V&V schedule are the key items to be considered. Among them, the evaluation criteria may include workload, correctness and accuracy, completeness, coincidence, traceability, error tolerance and so on. V&V methods mainly include static mock-up verification, dynamic mock-up verification shown as in Fig. 2, and full scope simulator-based simulation.

In the second phase (Evaluation), the V&V activities are implemented throughout the 3-phased control room design life cycle. For concept V&V, task analysis and operational review are the main V&V activities. For design V&V, HSI task support verification, HFE design verification [4], and checklist and traceability analysis are the main V&V activities. For integrated system V&V, process

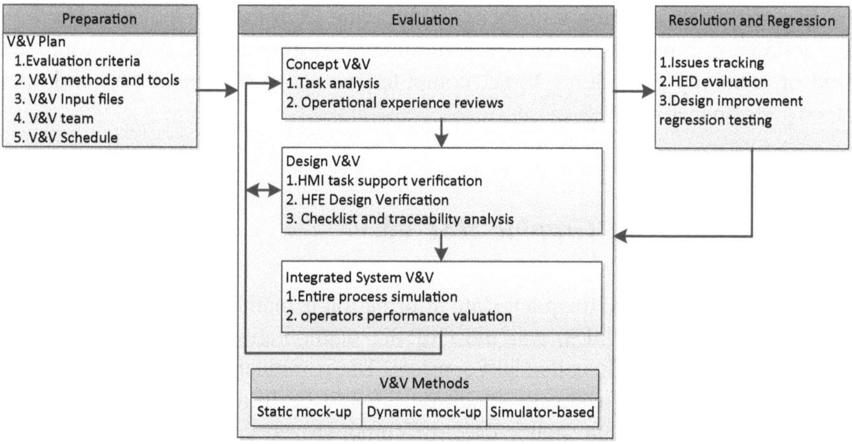

Fig. 1 V&V process and methods for digital control rooms

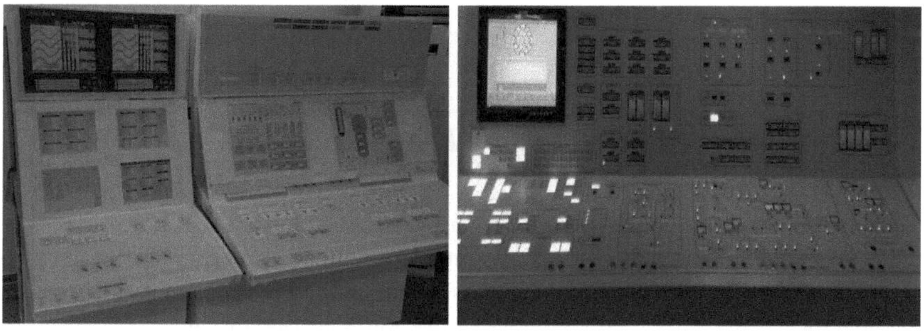

Fig. 2 Static mock-up and dynamic mock-up verification

simulation and operators performance validation are implemented. All of the activities are performed by three groups of operators with designers and suppliers.

The last phase (Resolution and Regression) is performed iteratively throughout V&V. The testing data are collected and the V&V issues (human engineering discrepancies, HED) are identified and resolved before the following V&V activities are started. Any design improvement should be verified by regression testing. Any residual issues should be analysed whether they do endanger the safety and performance of NPPs.

3 V&V Key Issues on Digital Control Room

The main characteristics of digital control rooms are the highly integrated computerized designs for display, control, alarm and procedures. How to run such computerized systems effectively, efficiently and safely is a key issue, which is still in discussion. This section will focus on the issue from the viewpoint of V&V for the following four aspects, including computerized graphic HSI V&V, computerized or hardcopy procedures V&V, computerized alarm systems V&V, and operators performance V&V.

3.1 Computerized Graphic HSI V&V

In digital control rooms, the plant state information is mainly displayed on the large screen displays and VDUs, and most of the manipulations are implemented by computer, which are also called soft controls. The computerized display and control graphs are not flat but stacked, which only could be browsed by means of page navigation. Therefore, the major tasks of computerized graphic HSI V&V are to inspect whether the designed graphs could support the operators' actions efficiently and effectively. Moreover, it should be checked whether there are errors or unnecessary items in the graphs, whether all required tasks are realized correctly, and whether the graphs are optimized.

3.2 Computerized or Hardcopy Procedures V&V

The main purpose of computerized or hardcopy procedures V&V is to verify that the HSI design and the procedures are coincident with each other and meet the plant safety and operational requirements. Usually, the hardcopy procedures are used for the operation of BUP (back up panel), and the computerized procedures are used for the computerized graphic HSI. In fact, although the digital technologies are used in control rooms, not all of the existing NPPs adopt the computerized procedures. For example, TIANWAN nuclear power plant is still using the hardcopy or paper procedures. Therefore, the following two cases should be considered for procedures V&V:

1. Computerized procedures V&V
2. Hardcopy or paper procedures V&V

For these two kinds of procedures, the common V&V tasks are to check and test the correctness of the procedures for the different operating conditions, including startup, operation, shutdown of the NPPs with normal, abnormal, alarm conditions, combating emergencies and other significant events. Especially, for the computerized procedures, it should be verified whether the contents of the procedures meet the requirements of the application of digital control rooms. For the hardcopy or paper procedures, it should be verified whether the actions listed in the procedures could be well performed by the computerized graphic displays and soft controls.

3.3 Computerized Alarm Systems V&V

Computerized alarm systems are one of the typical characteristics of the digital control rooms. On one hand, the computerized alarm systems substitute most of the traditional alarm systems. On the other hand, based on the digital technology, some new technologies, such as alarm suppression technology, are used to reduce the amount of alarms so as to avoid the excessive interference of alarms to operators and ensure the correctness of operators. Usually, there are two kinds of suppression technology. One is the alarm signal shielding technology used in Level 1. The other is alarm signal compression technology used in Level 2, where the alarm signals generated in the Level 1 are processed by the HSI using some constraint conditions. However, in order to maintain the higher safety of NPPs, some traditional alarm measures are still retained in control rooms.

Accordingly, the main tasks of computerized alarm systems V&V are to verify whether the computerized alarm systems and the reserved traditional alarm devices cooperate well with each other, and whether all alarms required by the system specification could be produced properly and timely.

3.4 Operators' Performance V&V

Operators are the diversified controllers of manipulating the NPPs, who are the critical backup of the automatic controllers and on whom the ultimate operational decisions are relied. The human factors failures could cause the safety risks. The operators' performance V&V is also the key issue during the life cycle of control rooms V&V. The main purpose of operators' performance V&V is to verify whether they could accomplish their mission efficiently and correctly under any anticipated conditions. In practice, the V&V range of process simulation is constrained by the V&V schedule and V&V human resources. Partial process simulation and validation is usually implemented under some typical operating conditions, such as startup, operation, shutdown of the NPPs with normal and abnormal conditions, by which the operators' performance are verified.

4 Application Practice

In this section, some example application practices are given to show the effectiveness of digital control rooms V&V.

1. Optimized computerized graphic HSI

The early computerized graphic HSI with two graphs and the optimized HSI with four graphs for reactor coolant system (RCS) are shown in Figs. 3 and 4

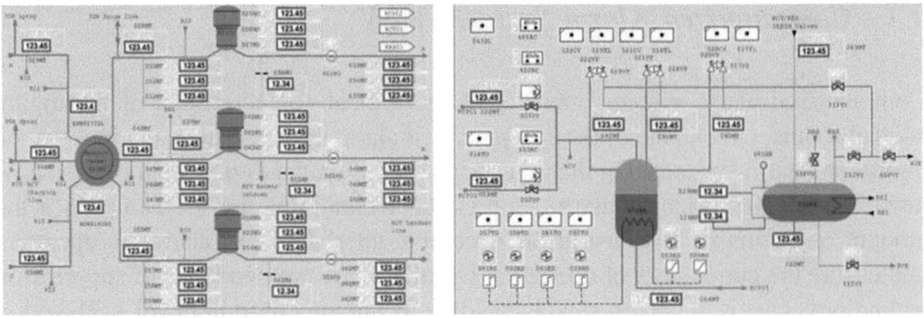

Fig. 3 Early computerized graphic HSI design for RCS

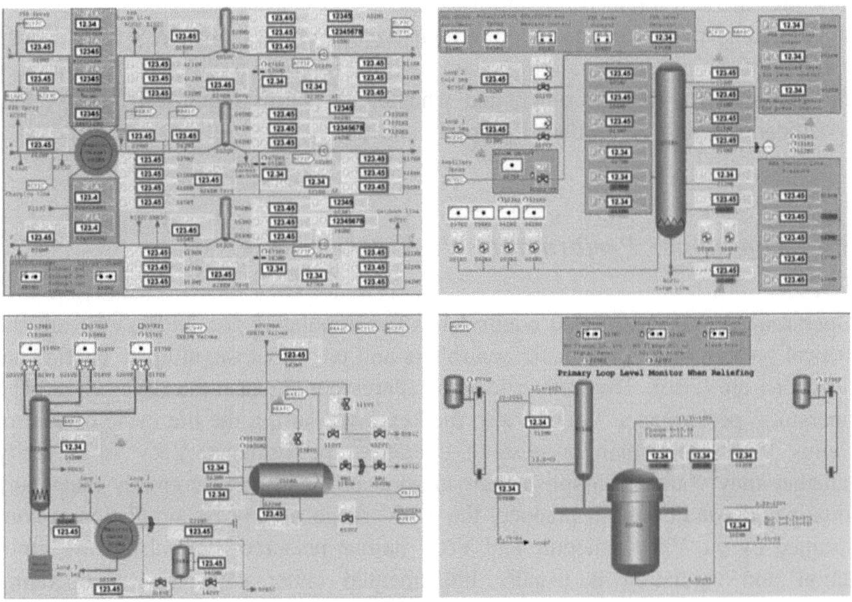

Fig. 4 Optimized computerized graphic HSI design for RCS

respectively. It is shown by Fig. 4 that more information for the each main equipment in RCS are accommodated and laid out properly.

2. Optimized operator workstation

As shown in Fig. 5, the left figure is the early covered-type operator workstations and the right figure is the optimized bracket-type ones. By such change, it is easy to install the operator workstations, and to adjust the place and the visual angle of operator workstations as well.

Fig. 5 Covered-type and bracket-type operator workstation

5 Conclusion

With the digital technologies are used in the control rooms of NPPs, the information displays, controls and alarm systems are changed into the highly integrated computerized design. Most of the manipulations are accomplish through page navigation. How to effectively and efficiently maintain the plant's safety and performance is a critical topic for the design and V&V technologists. From the viewpoint of V&V, the paper concludes and analyzes some important issues during the life cycle of digital control room V&V to ensure the quality of control rooms.

References

1. USNRC, Human-System Interface Design Review Guidelines, NUREG 0700, Rev 2, 2002.
2. USNRC, Human Factors Engineering Program Review Models, NUREG 0711, 2004.
3. Zhang J.B., Sun Y.B., Liu Y.Z., Study on Human Factor Engineering V&V method System of Advanced Main Control Room in Nuclear plant (In Chinese). Electronic Instrumentation, 20 (5): 23–27, 2013.
4. Liang K., Kang L.H., Feng W., Application Research of HFE V&V Based on the Design of Nuclear Power Control Room (In Chinese). Computer Knowledge and Technology, 9(20): 4752–4757.

Research About the Spurious Trip Rate of Digital RPS Based on FMEA

Jun-Yi Li, Hui-Hui Liang and Ya-Nan He

Abstract Digital control system (DCS) is the central nervous system of automation monitoring and controlling of nuclear power plant. Safe operation and economic benefits of nuclear reactor units depend heavily on the performance of DCS. DCS consists of hardware and control logic. Mal-Operation Rate and Anti-Operation Rate are two important measures to evaluate the reliability of DCS. Failure Mode and Effects Analysis (FMEA) is now in common use to analyze the failures of hardware and control logic. FMEA has been widely used in assessing DCS of nuclear power plant. Fault Tree Analysis (FTA) is effective in analyzing causes and effects of failures. Based on the top-down analyzing strategy, it can easily identify the basic reason accident. Reactor protection system (RPS) is the most important part of DCS, which is used to guarantee the reactor stay in a safety state. This paper takes Spurious Trip Rate of RPS as example, analyzing how different failure modes affect reliability of RPS based on the method of FMEA and FTA. This method can also apply to analyze engineered safety features actuation system (ESFAS), and even the whole DCS of nuclear power plant, which can provide guidance for the design, operation and maintenance of nuclear power plant.

Keywords DCS · RPS · FMEA · FTA · Spurious trip rate

1 Introduction

RPS is an important safety system of nuclear power plant. It will be triggered automatically when the system encounter failures. Nuclear power plants always use FMEA to identify potential failures which may affect the reliability of RPS. It is a systematic process for identifying potential failures before they occur, with the

J.-Y. Li (✉)
Rensselaer Polytechnic Institute, Troy, USA
e-mail: lijunyi0819@gmail.com

H.-H. Liang · Y.-N. He
China Nuclear Power Design CO., LTD, Shenzhen, China

Y. Xu (ed.), *Nuclear Power Plants: Innovative Technologies for Instrumentation and Control Systems*, Lecture Notes in Electrical Engineering 400,
DOI 10.1007/978-981-10-3361-2_24

intent to eliminate or minimize the risks associated with them. It allows validating the architecture of the protection system with regards to the Single Failure criterion.

The FMEA for safety functions which are implemented in protection system only presents the failure effects and consequences on the functions when the system encounters single failure. It only orients toward qualitative analysis on the failure mode of hardware: safe or unsafe and what is the functional consequences. When it is unsafe, Reactor Trip System and Engineered Safety Features Actuation System will be triggered. However, when assessing the reliability of the software, quantitative analysis should be brought into the evaluation. In this way, the safety functions which are implemented in protection system could be assessed and improved after the quantitative analysis. According to the reliability level of the system, engineers could make adaptive changes to upstream design in order to ensure normal function of power plants.

Fault Tree Analysis (FTA) is one of the most important logic and probabilistic techniques used in system safety and reliability assessment today. In order to quantify their contribution to system unreliability in the course of product design, it investigates potential faults and their modes and causes. Because of its logical, systematic and comprehensive approach, FTA has been proven capable of uncovering design and operational weaknesses that escaped even some of the best deterministic safety and engineering experts.

2 RPS Failure Modes

Reactor Spurious Trip occurs when Reactor Trip Actuating Signal encounters spurious trip and the Reactor Trip Breaker is triggered. Reactor Trip Breaker is made up of hardware, while Reactor Trip Actuating Signal consists of hardware and software. Here the paper is discussing the reliability of Reactor Trip Actuating Signal.

Reactor Trip Actuating Signal is produced from four Reactor Protection Cabinet (RPC) channels through 2-out-of-4 voting logic. Each RPC channel is composed of input part, communication part, processing part and output part. Input part collects signals in the channel and communication part collects signals from other channels. Then processing part processes the signals through logical operations and output part outputs Reactor Trip Actuating Signal from the channel. Finally Reactor Trip Breaker is triggered. Once one of the four parts encounters failures would lead to channel failure and then spurious trip.

Assume the possibilities of each part encounters failures are P1, P2, P3 and P4. Thus the possibility Reactor Trip Actuating Signal trigger spurious trip consists of P1, P2, P3 and P4. Manipulating Fault Tree Analysis and the functional consequences implemented by [1] the Protection System, this paper deduces the expression of the Spurious Trip Rate of Reactor Trip Actuating Signal and checks which failure affects the whole system the most (Fig. 1).

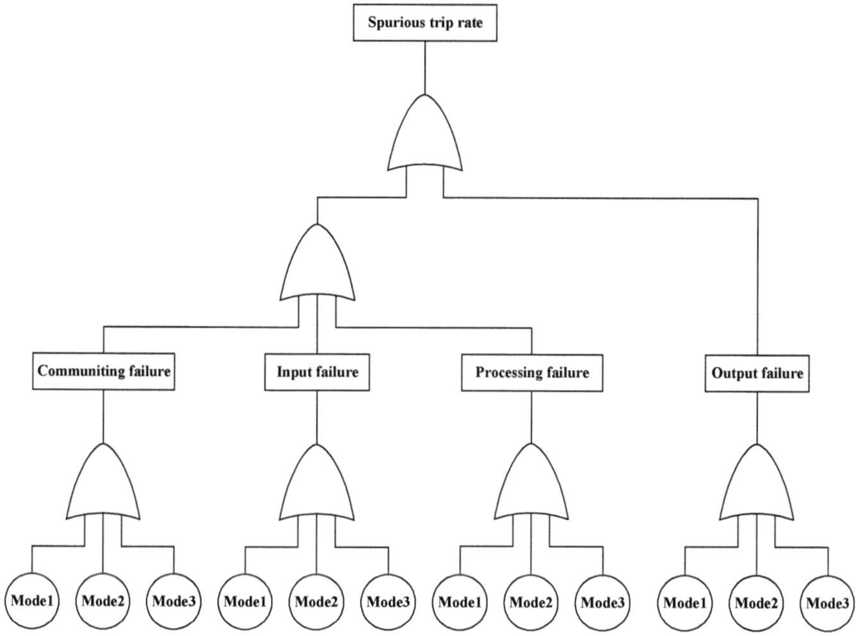

Fig. 1 RPS spurious trip rate diagram

Table 1 Occurrence rate of three failure modes

Failure mode	Description	Occurrence rate
Mode 1	Detected by self-diagnosis	R
Mode 2	Undetected and actuated	S
Mode 3	Undetected and not actuated	T

Taking the input part of RPS as example, this paper analyzes the effects on Spurious Trip Rate from single failure. Failures generated from RPS input part are divided into three modes according to if they could be detected:

Mode 1: failures detected by self-diagnosis;
Mode 2: failure undetected by self-diagnosis and resulting in the spurious protective actuation;
Mode 3: failure undetected by self-diagnosis and resulting in no protective actuation.

Assume each failure mode has an occurrence rate as listed in Table 1.

3 RPS Voting Logic Degrade

RPS consists of four channels and Reactor Trip Actuating Signal is produced after the output signal being processed by voting logic 2-out-of-4 in the channels. Ignoring bypassing situations, voting logic would be degraded to 2-out-of-3 when one of channels encounters failure.

If one of the other channels encounters failure at this time, the voting logic will be degraded to 1-out-of-2. When all three channels encounter failures, the voting logic would be degraded to 1-out-of-1. Namely, Reactor Trip Actuating Signal from the other one channel must trigger reactor trip. Due to undetectable failures in the system, there are voting logic 1-out-of-3, 2-out-of-2 and 1-out-of-1. Here the paper assumes each situation has an occurrence rate of A, B and C separately [2].

The system is taking voting logic 2-out-of-4 when the reactor in normal running. When any channel encounters failure, the voting logic would be degraded. Three failure effects and functional consequences of the voting logic 2-out-of-4 are shown in Table 2.

Table 2 2-out-of-4 voting logical degrade process

Failure mode	Failure effect	Functional consequences
Mode 1	Logic is degraded	Reactor Trip Logic (RTL) becomes 2-out-of-3 due to the logic degradation. Remaining 3 channels provide reactor trip. If unrestricted by pass of one instrument channel has already been executed in another channel, RTL becomes 1-out-of-2 due to the further logic degradation. Remaining 2 channels provide reactor trip
Mode 2	Bistable changes to trip state and partial trip signal is generated in the affected RPC channel	RTL becomes 1-out-of-3 due to the sensor failure. Remaining 3 channels provide reactor trip. If unrestricted by pass of one instrument channel has already been executed in another channel, RTL becomes 1-out-of-2 due to the further logic degradation. Remaining 2 channels provide reactor trip
Mode 3	Bistable does not change to trip state in the affected RPC channel when process reaches trip level	RTL becomes 2-out-of-3 due to the input failure. Remaining 3 channels provide reactor trip. If unrestricted by pass of one instrument channel has already been executed in another channel, RTL becomes 2-out-of-2 due to the further logic degradation. Remaining 2 channels provide reactor trip

Table 3 2-out-of-3 voting logical degrade process

Failure mode	Failure effect	Functional consequences
Mode 1	Logic is degraded	RTL becomes 1-out-of-2 due to the logic degradation. Remaining 3 channels provide reactor trip. If unrestricted by pass of one instrument channel has already been executed in another channel, RTL actuates due to the further logic degradation
Mode 2	Bistable changes to trip state and partial trip signal is generated in the affected RPC channel	RTL becomes 1-out-of-2 due to the sensor failure. Remaining 3 channels provide reactor trip. If unrestricted by pass of one instrument channel has already been executed in another channel, RTL actuates due to the sensor failure
Mode 3	Bistable does not change to trip state in the affected RPC channel when process reaches trip level	RTL becomes 2-out-of-2 due to the input failure. Remaining 3 channels provide reactor trip. If unrestricted by pass of one instrument channel has already been executed in another channel, RTL becomes 1-out-of-1 due to the sensor failure. Remaining 2 channels provide reactor trip with 1-out-of-1 logic

When voting logic 2-out-of-4 degraded to 2-out-of-3, if any other channel encounters failure, then the voting logic would be degraded furthermore. Three failure effects and functional consequences of the voting logic 2-out-of-3 are shown in Table 3.

When voting logic 2-out-of-3 degraded to 1-out-of-2, if any other channel encounters failure, then the voting logic would be degraded furthermore. Three failure effects and functional consequences of the voting logic 1-out-of-2 are shown in Table 4.

According to the analysis results of Tables 2, 3 and 4, the degrade process of reactor protection system voting logic can be shown as Fig. 2.

4 RPS Spurious Trip Rate

When the system encounters these three failure modes, the voting logic would be degraded. Spurious Trip Rate at this time is the product of occurrence rate of the failure mode and spurious probability of voting logic for next state, which is Spurious Trip Rate = (spurious trip probability of first state) × (spurious trip probability of second state) [3].

Table 4 1-out-of-2 voting logical degrade process

Failure mode	Failure effect	Functional consequences
Mode 1	Logic is degraded	RTL becomes 1-out-of-1 due to the logic degradation. Remaining 3 channels provide reactor trip
Mode 2	Bistable changes to trip state and partial trip signal is generated in the affected RPC channel	RTL actuates due to the sensor failure
Mode 3	Bistable does not change to trip state in the affected RPC channel when process reaches trip level	RTL becomes 1-out-of-1 due to the input failure. Remaining 3 channels provide reactor trip with 1-out-of-1 logic

Fig. 2 RPS voting logic degrade process

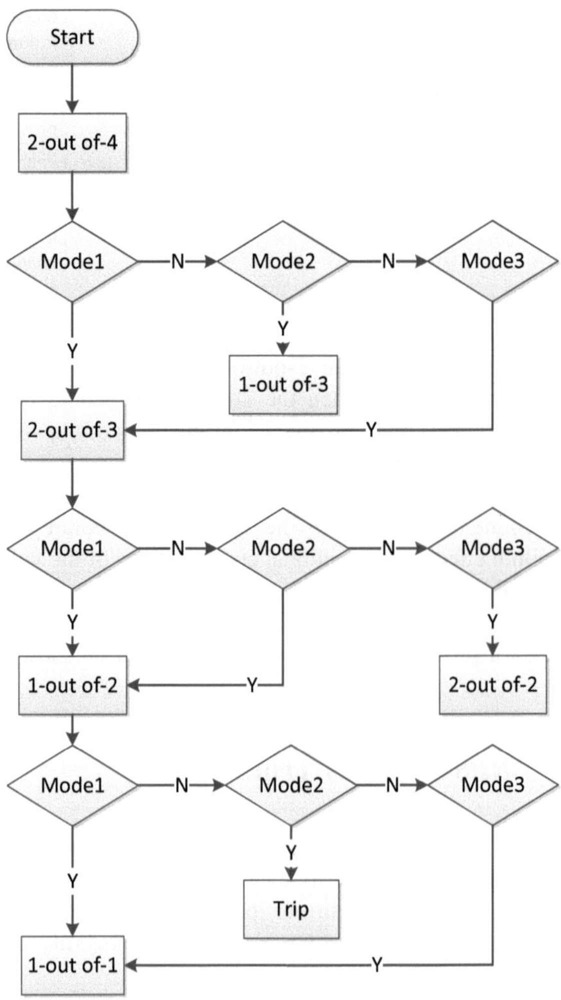

Table 5 1-out-of-2 trip function spurious trip rate

Failure mode	Second state	Spurious trip rate
Mode 1	1-out-of-1 logic	RA
Mode 2	If sensor fails → Reactor trip	0
Mode 3	1-out-of-1 logic	TA

Table 6 2-out-of-3 trip function spurious trip rate

Failure mode	Second state	Spurious trip rate
Mode 1	1-out-of-2 logic	R(RA + TA)
Mode 2	1-out-of-2 logic	S(RA + TA)
Mode 3	2-out-of-2 logic	TB

Table 7 2-out-of-4 trip function spurious trip rate

Failure mode	Second state	Spurious trip rate
Mode 1	2-out-of-3	R(S + R) (RA + TA) + RTB
Mode 2	1-out-of-3	SC
Mode 3	2-out-of-3	$T(S + R)(RA + TA) + T^2B$

According to Table 4, when the system is taking voting logic 1-out-of-2 and encountering one of the failure modes, it would trigger reactor trip or the voting logic would be degraded (becoming 1-out-of-1). Spurious Trip Rate of these different situations is shown in Table 5. For spurious trip caused by any channel failure, 1-out-of-2 Trip Function has a Reactor Spurious Trip Rate: (RA + TA).

According to Table 3, when the system is taking voting logic 2-out-of-3 and encountering one of the failure modes, the voting logic would be degraded, becoming 1-out-of-2 or 2-out-of-2. Combining the results from Table 5, Spurious Trip Rate of these different situations is shown in Table 6. For spurious trip caused by any channel failure, 2-out-of-3 Trip Function has a Reactor Spurious Trip Rate: R(RA + TA) + S(RA + TA) + TB.

According to Table 2, when the system is taking voting logic 2-out-of-4 and encountering one of the failure modes, the voting logic would be degraded, becoming 1-out-of-3 or 2-out-of-3. Combining the results from Table 6, Spurious Trip Rate of these different situations is shown in Table 7. For spurious trip caused by any channel failure, 2-out-of-4 Trip Function has a Reactor Spurious Trip Rate: $(R + T)[(S + R)(RA + TA) + TB] + SC = R^3A + AR^2(2T + S) + R(AT^2 + 2ATS + TB) + T^2(AS + B) + SC$.

For 2-out-of-4 Trip Function, probability of occurrence of Failure Mode 1 affects the spurious trip rate the most. As results, engineers should pay more attention to this kind of failure in order to achieve proper operation.

5 Conclusions

This paper takes RPS as example, which focusing on the input part of RPS. Analyzing how different failure modes affect reactor spurious trip using the method of FMEA and FTA. Getting the spurious rate of nuclear reactor through calculating the spurious trip rate of 2-out-of-4, 2-out-of-3, 1-out-of-2, then identify the important part of RPS, which can use to provide guidance for the design, operation and maintenance of nuclear power plant. This research method can also be applied to the analysis of processing, communication and output part of RPC. Furthermore, engineers can generalize this analysis method to evaluate the reliability of the whole DCS.

References

1. Hong-Wei Li, Sun Yu. Analysis of Shutdown Operation Reliability of Reactor Trip Breaker. Nuclear Power Engineering. 2013(34): 169–171.
2. Jia-Sheng Wang, Zhen-Min Hong, Ping Hu. Some digit I&C system applies and analyze in nuclear power station's reconstruct. Chinese Journal of Nuclear Science and Engineering, 2005-03.
3. Galyean W J. Digital control systems in nuclear power plants: Failure information, modeling concepts, and applications. Revision 1[J], 1993.

MIX
Papier aus verantwortungsvollen Quellen
Paper from responsible sources
FSC® C105338

If you have any concerns about our products,
you can contact us on
ProductSafety@springernature.com

In case Publisher is established outside the EU,
the EU authorized representative is:
Springer Nature Customer Service Center GmbH
Europaplatz 3, 69115 Heidelberg, Germany

Printed by Libri Plureos GmbH
in Hamburg, Germany